Spring Boot 开发实战

微课视频版

吴胜 ◎ 编著

清華大學出版社
北京

内 容 简 介

本书先介绍 Spring Boot 开发的基础知识，包括 Spring Boot 简介、Spring Boot 开发起步、Spring Boot 的相关注解、Spring Boot 的 Web 应用开发、Spring Boot 的数据库访问、Spring Boot 的 Web 服务开发、Spring Boot 的数据处理、Spring Boot 的文件应用、Spring Boot 的 WebFlux 开发，然后结合一个简单案例演示了 Spring Boot 开发的全过程。本书结合示例由浅入深地介绍各个知识点，并按开发步骤组织各个章节的内容，同时配备了示例的操作视频，可以帮助读者更好地理解、掌握 Spring Boot 开发技术。

本书内容通俗易懂，适合 Spring Boot 开发的初学者（特别是在校学生）、Web 应用开发者和企业级应用开发爱好者等学习，也可以作为高等学校的教材、自学的入门读物、开发过程的参考书。

图书在版编目（CIP）数据

Spring Boot 开发实战：微课视频版 / 吴胜编著. —北京：清华大学出版社，2019(2024.2重印)
（清华科技大讲堂）
ISBN 978-7-302-52819-7

Ⅰ. ①S… Ⅱ. ①吴… Ⅲ. ①JAVA 语言-程序设计 Ⅳ. ①TP312.8

中国版本图书馆 CIP 数据核字（2019）第 082670 号

责任编辑：陈景辉　张爱华
封面设计：刘　键
责任校对：焦丽丽
责任印制：沈　露

出版发行：清华大学出版社
网　　址：https://www.tup.com.cn，https://www.wqxuetang.com
地　　址：北京清华大学学研大厦 A 座　　邮　　编：100084
社 总 机：010-83470000　　邮　　购：010-62786544
投稿与读者服务：010-62776969，c-service@tup.tsinghua.edu.cn
质 量 反 馈：010-62772015，zhiliang@tup.tsinghua.edu.cn
印 装 者：三河市铭诚印务有限公司
经　　销：全国新华书店
开　　本：185mm×260mm　　**印　　张**：20.75　　**字　　数**：501 千字
版　　次：2019 年 9 月第 1 版　　**印　　次**：2024 年 2 月第 6 次印刷
印　　数：6301～6800
定　　价：59.90 元

产品编号：077866-01

前 言

Spring 以简化企业级应用开发为己任。无论是 Web 应用开发、数据库访问还是当前的大数据处理、分布式应用集成，都能看到 Spring 的身影。然而，一转眼，作为 EJB 颠覆者的 Spring 也从最初的轻量级工具变成了"庞然大物"。而 Spring Boot 由于能极大地简化配置，并且能和当下流行的微服务架构契合，因此一出现便受到了大家的追捧。

Spring Boot 在 Java 应用开发领域快速兴起，其原因除了它具有约定大于配置、采用更简洁的配置方式来替代 XML 等特点外，还有一个重要原因是用 Spring Boot 来开发时不需要同时面对多个框架（如 Struts2、Spring 和 Spring MVC、Hinernate 或 MyBatis）和不同的视图显示技术（如 JSP、Servlet 等）。

不同框架之间的联系、整合问题以及由此带来的更加复杂的配置问题（特别是利用 XML 进行配置时）是 Spring 学习者在学习时（特别是入门阶段）需要面对的一个重要难题。而 Spring Boot 较好地封装了相关工具和框架（如 Tomcat、Hibernate、MySQL 驱动包等），可以开箱即用这些工具和框架，使得 Spring Boot 开发比较简单。

但是，软件开发领域"没有银弹"。Spring Boot 全面封装、开箱即用使得开发变得更加快速、透明、高效的同时，对于初学者而言，Spring Boot 开发时的依赖管理和配置信息设置问题是需要面对的一个挑战。好在开发工具（如 Spring Tool Suite、IntelliJ IDEA）以及帮助文档可以有效地帮助 Spring Boot 初学者降低学习难度。

另外，Spring Boot 的应用比较多，这使得 Spring Boot 的内容略显庞杂。而且，Spring Boot 还在快速地更新，这会导致本书介绍的一些知识点在新版本中可能会有更新，于是需要读者在开发时参考官方文档进行知识更新。这些因素也增加了 Spring Boot 的学习难度。

为了帮助读者更好地掌握 Spring Boot 开发技术，本书按照开发步骤组织各章节的内容，循序渐进地介绍 Spring Boot 的开发知识和示例代码。为了帮助读者更好地安排学习时间和帮助教师更好地安排授课，在下表中给出了各章的建议学时（建议学时分为建议理论学时和建议实践学时）。

章　内　容	建议理论学时	建议实践学时
第 1 章　Spring Boot 简介	2	0
第 2 章　Spring Boot 开发起步	2	2
第 3 章　Spring Boot 的相关注解	3	0
第 4 章　Spring Boot 的 Web 应用开发	5	3
第 5 章　Spring Boot 的数据库访问	5	3
第 6 章　Spring Boot 的 Web 服务开发	5	3
第 7 章　Spring Boot 的数据处理	4	2
第 8 章　Spring Boot 的文件应用	4	2
第 9 章　Spring Boot 的 WebFlux 开发	4	3
第 10 章　Spring Boot 开发案例	2	2
合计学时	36	20

在开设 Spring Boot 开发的相关课程时可以根据总学时、学生基础和教学目标等情况调整各章的学时。学习者也可以有选择地阅读各章节内容并安排好学时。

为便于教学，本书有教学视频、源代码、课件等配套资源。

（1）获取教学视频方式：读者可以先扫描本书封底的文泉云盘防盗码，再扫描书中相应的视频二维码，观看教学视频。

（2）获取源代码及参考答案方式：先扫描本书封底的文泉云盘防盗码，再扫描下方二维码，即可获取。

源代码及参考答案

源代码使用说明

（3）其他配套资源可以扫描本书封底的课件二维码下载。

由于时间短，加上编者水平有限，书中难免有疏漏之处，敬请读者朋友批评指正。

编　者

2019 年 1 月

目录

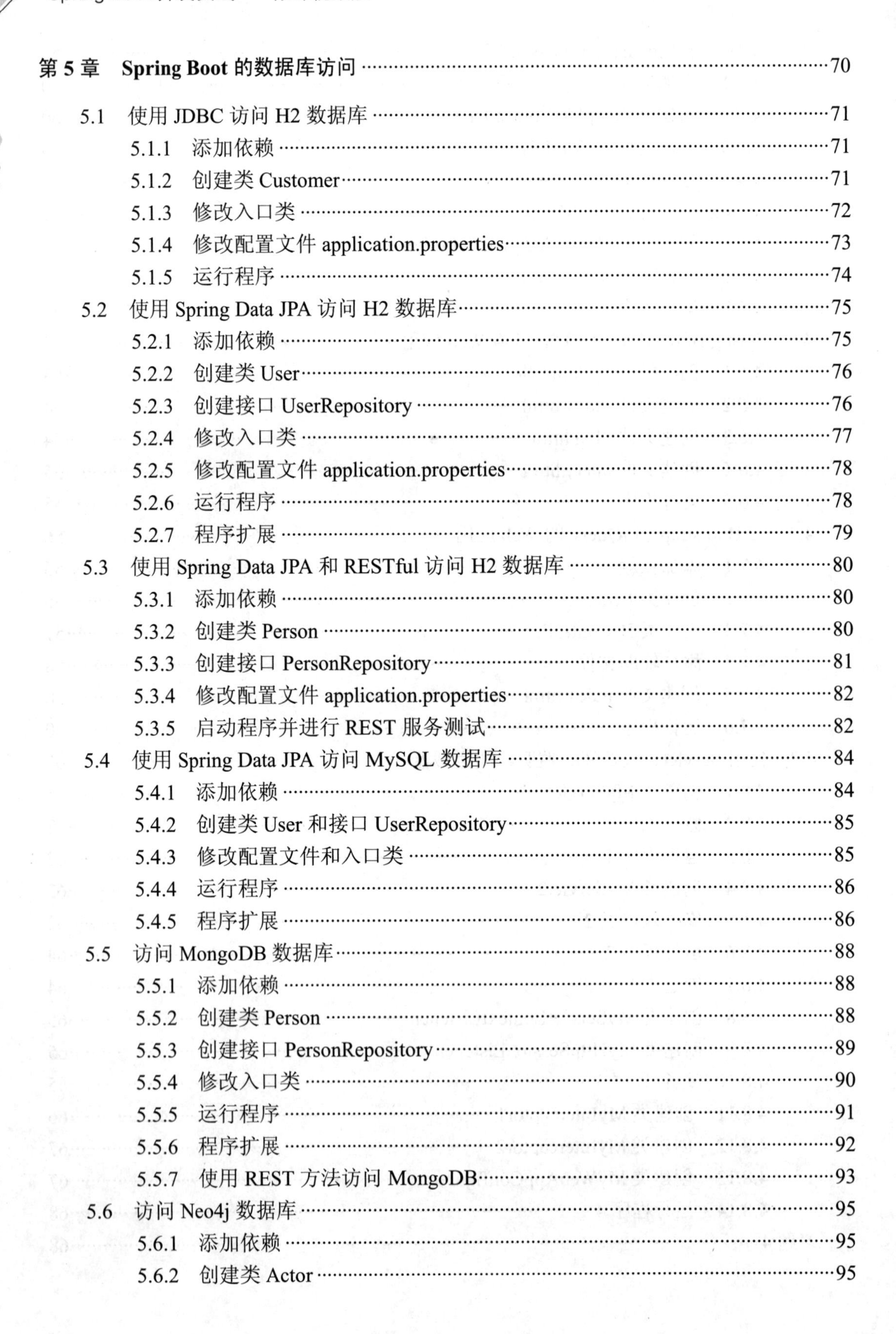

第 1 章

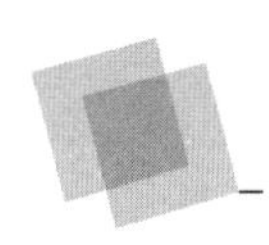

Spring Boot 简介

本章主要介绍 Spring Boot 的发展背景、Spring Boot 的特征、Spring Boot 的工作机制等内容。

1.1 Spring Boot 的发展背景

1.1.1 Spring 的发展

作为 EJB 的颠覆者，Spring 因为轻量级的开发方式很快被业界接受。Spring 的发展使得它也逐步变成了“庞然大物”。历史有惊人的相似，Spring Boot 再次因其轻量级的开发方式而受到大家的追捧。

回顾 Spring 的发展过程可以发现：Spring 1.X 时代是通过 XML 文件配置 Bean 并实现 Spring 与其他框架的集成。随着项目的扩大，配置 XML 过于烦琐，而且 XML 文件变得更加臃肿不堪，难以理解、设计、配置和管理。Spring 2.X 时代可以通过 Java 5 带来的注解对 Bean 进行声明注入。从 Spring 3 开始，Spring 使用 Java 配置方式来管理 Bean，从而大大地减少了 XML 的使用，甚至是零配置。

Spring 框架由二十多个模块组成，可分为 Test（测试）、Core Container（核心容器）、AOP（面向切面编程）、Aspects（切面）、Instrumentation（监测仪）、Messaging（消息）、Data Access/Integration（数据访问/集成）、Web 等模块，结构如图 1-1 所示。其中，核心容器包括 Beans（模块名为 spring-beans）、Core（模块名为 spring-core）、Context（模块名为 spring-context）和 SpEL（Spring Expression Language，Spring 表达式语言，模块名为 spring-expression）等模块；数据访问/集成模块包括 JDBC（模块名为 spring-jdbc)、ORM（对象关系映射，模块名为 spring-orm）、OXM（对象 XML 映射，模块名为 spring-oxm）、

JMS（Java 消息服务，模块名为 spring-jms）、Transactions（事务，模块名为 spring-tx）等模块；Web 模块包括 WebSocket（模块名为 spring-websocket）、Servlet（模块名为 spring-webmvc）、Web（模块名为 spring-web）、Portlet（模块名为 spring-webmvc-portlet）等模块。

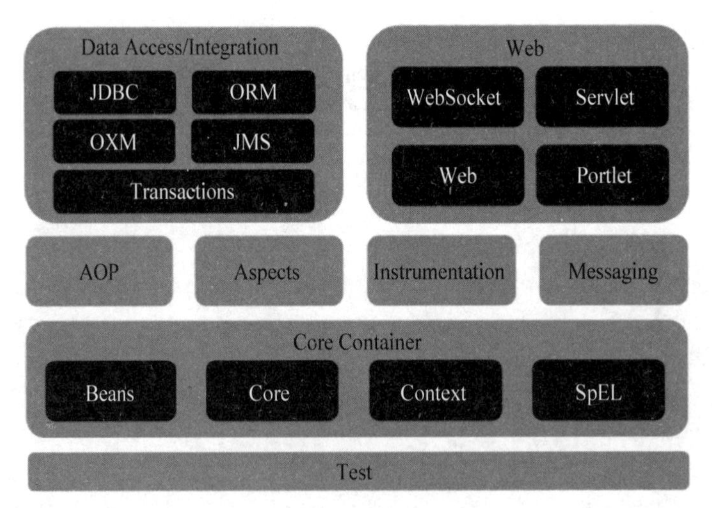

图 1-1 构成 Spring 框架的模块图

1.1.2 Spring 的生态圈

Spring Boot 是伴随 Spring 4 而诞生的。除了 Spring Boot 之外，Spring 作为企业应用开发的轻量级解决方案提供了许多子项目；了解这些子项目可以更好地理解其设计架构、思想并使用 Spring。Spring 的整个生态系统包括以下内容。

Spring Framework(Core)：Spring 的核心项目，其中包含了一系列 IoC 容器的设计，提供了依赖注入的实现；同时，还集成了 AOP，提供了面向切面编程的实现；当然还有 MVC、JDBC、事务处理模块的实现。

Spring Boot：提供了快速构建 Spring 应用的解决方案，达到“开箱即用”；使用默认的 Java 配置来实现快速开发，并“即时运行”。

Spring Batch：提供构建批处理应用和自动化操作的框架，专门用于离线分析程序、数据批处理等场景。

Spring Data：对主流的关系型数据库提供支持，并提供使用非关系型数据的能力，如将数据存储在非关系数据库或 MapReduce 中的分布式存储、云计算存储环境等。

Spring Security：通过用户认证、授权、安全服务等工具保护应用，它最先在 Spring 社区中的名字是 Acegi 框架。

Spring Security OAuth：OAuth 是一个第三方的模块，提供了一个开放的协议的实现，通过这个协议前端应用可以对 Web 应用进行简单而标准的安全调用。

Spring Web Flow：基于 Spring MVC 提供 Web 应用开发。它是 Web 工作流引擎，定义了一种特定的语言来描述工作流；同时高级的工作流控制器引擎可以管理会话状态。

Spring BlazeDS Integration：提供 Spring 与 Adobe Flex 技术集成的模块。

Spring Dynamic Modules：提供 Spring 运行在 OSGi 平台上面向 Java 的动态模型系统，Eclipse 就是构建在 OSGi 平台上的。

Spring Integration：通过消息机制为企业的数据集成提供了解决方案。

Spring AMQP：高级消息队列协议（Advanced Message Queuing Protocol），支持 Java 和.NET 两个版本。AMQP 是一个提供统一消息服务的应用层标准高级消息队列协议，是一个开放标准，为面向消息的中间件设计，如 Rabbit MQ 等。SpringSoruce 旗下的 Rabbit MQ 就是一个开源的 AMQP 的消息服务器，Rabbit MQ 是用 Erlang 语言开发的。

Spring .NET：为.NET 提供与 Spring 相关的技术支持，如 IoC 容器、AOP 等。

Spring for Android：为 Android 终端开发应用提供 Spring 支持。

Spring Mobile：为移动终端的服务器应用开发提供支持。

Spring Social：Spring 框架的扩展，提供与社交网 SNS 服务 API（如 Facebook、新浪微博和 Twitter 等）的集成。

Spring XD：用来简化大数据应用开发。

Spring Cloud：为分布式系统开发提供工具集。

Spring HATEOAS：基于 HATEOAS 原则简化 REST 服务开发。HATEOAS 是“超文本驱动”（Hypermedia As The Engine Of Application State）的英文缩写，又名“将超媒体作为应用状态的引擎”。

Spring Web Services：提供了基于协议有限的 SOAP/Web 服务。SOAP 是简单对象访问协议（Simple Object Access Protocol）的英文缩写。

Spring LDAP：简化使用 LDAP 开发。LDAP 是轻量级目录访问协议（Lightweight Directory Access Protocol）的英文缩写。

Spring Session：提供一个 API 及实现来管理用户会话信息。

1.1.3 Spring Boot 的发展

Spring Boot 是伴随 Spring 4 而诞生的，它在继承 Spring 优点的基础上，简化了基于 Spring 的开发。Spring Boot 使得开发者可以更容易地创建基于 Spring 的可以“即时运行”的应用和服务。所以，一经推出就引起业界极大的关注。

Spring Boot 是由 Pivotal 团队开发的，其设计目的是简化创建 Spring 应用的初始搭建和开发过程。Spring Boot 使用特定的方式进行配置，使得开发人员不再需要定义样板化的配置。通过这种方式 Spring Boot 成为了快速应用开发（Rapid Application Development, RAD）领域的领导者。

Spring Boot 并不提供 Spring 框架的扩展功能，只是用于快速地开发基于 Spring 的应用程序。它并不是 Spring 的替代物，而是和 Spring 紧密结合来提高 Spring 应用开发者开发体验的工具。

2014 年 4 月发布了 Spring Boot 1.0。本书开始编写时，Spring Boot 最新版本为 2.0.2 版；本书编写完成时最新版本为 2.0.6 版。本书的例子主要基于 Spring Boot 2.0.2 编写，例子中一直用的是当时最新版本，个别例子用到了 2.0.6 版，但是这些版本的不同不会影响到本书的使用。

1.2 Spring Boot 的特征

1.2.1 Spring Boot 的特点

Spring Boot 使得创建独立（或产品级）的基于 Spring 应用变得更容易，大多数 Spring Boot 应用只需要很少的 Spring 配置。开发 Spring Boot 的主要目的是为所有 Spring 开发者提供一个更迅速、可用的入门经验。Spring Boot 坚持“开箱即用”，为具有许多类的工程提供一系列常用的非功能特性（例如嵌入式服务器、安全、度量、健康检查、外部配置等）。

Spring Boot 的特点如下。

（1）约定大于配置。通过代码结构、注解的约定和命名规范等方式来减少配置，并采用更加简洁的配置方式来替代 XML 配置；减少冗余代码和强制的 XML 配置。

（2）能创建基于 Spring 框架的独立应用程序。

（3）内嵌有 Tomcat，无须部署 War 文件。

（4）简化 Maven 配置，并推荐使用 Gradle 替代 Maven 进行项目管理。Maven 用于项目的构建，主要可以对依赖包进行管理。Maven 将项目所使用的依赖包信息放到 pom.xml 文件的<dependencies></dependencies>结点之间。

（5）自动配置 Spring。

（6）提供生产就绪型功能。提供了一些大型项目中常见的非功能特性，如嵌入式服务器、安全、指标、健康检测、外部配置等内容。

（7）定制“开箱即用”的 Starter，没有代码生成，也无须 XML 配置；还可以修改默认值来满足特定的需求。

（8）为 Spring 开发者提供更快速的入门体验。Spring Boot 不是对 Spring 进行功能上的增强，而是提供了一种更快速的 Spring 使用方法。

（9）对主流框架无配置集成，自动整合第三方框架，如 Struts。

（10）使用注解使编码变得更加简单。

（11）与基于 Spring Cloud 的微服务开发无缝结合。

1.2.2 Spring Boot 2 的新特性

相对于 Spring Boot 1，Spring Boot 2 是一个重要的更新版本，它新增的特性有以下三个方面。

（1）对 Gradle 插件进行了重写，以便于项目管理。

（2）基于最新的 Java 8 和 Spring Framework 5。Java 8 中引入了函数式编程，并极大地改善了并发程序开发体验；Spring Framework 5 推出了新的响应式 Web 框架，这使得基于 Spring Boot 2 开发企业级应用变得更加简单，可以更方便地构建响应式编程模型。

（3）对与 Spring Boot 相关的 Spring Data、Spring Security、Spring Integration 等内容进行了更新。

由于本书所用到的版本是 2.0.2 及以后版本，所以下文除非是为了区分或强调，否则 Spring Boot 均指 Spring Boot 2。

1.2.3 Spring Boot 2 的核心模块

Spring Boot 2 中 spring-boot 模块是 Spring Boot 2 的核心工程。

spring-boot 模块提供了一些特性来支持 Spring Boot 的其他模块，这些特性包括以下四个方面。

（1）SpringApplication 类提供了静态方法以便于写一个独立的 Spring 应用程序，该类的主要职责是创建和更新一个合适的 Spring 应用程序上下文（ApplicationContext）。

（2）给 Web 应用提供了一个可选的 Web 容器（如 Tomcat 或 Jetty 等）。

（3）通过 application.properties 文件等方式提供一流的外部配置的支持。

（4）提供了便捷的应用程序上下文的初始化器，以便在使用它之前对其进行用户定制。

Spring Boot 中 spring-boot-autoconfigure 模块是实现自动配置（auto-configuration）的核心工程。Spring Boot 可以依据 classpath 里依赖的内容来自动配置 Bean 到 IoC 容器，但是要开启这个自动配置功能需要添加@EnableAutoConfiguration 注解。auto-configuration 会尝试推断哪些 Bean 是用户可能会需要的。例如，在当前 classpath 下有 HSQLDB 包，并且用户没有配置其他数据库链接，这时自动配置功能会自动注入一个基于内存的数据库连接到应用的 IoC 容器。目前 auto-configuration 提供 Web、JDBC、Spring Data JPA、Spring Batch、Thymeleaf、Reactor 等注解。auto-configuration 使用在 class 上标注@Configuration 注解实现。使用@Configuration 时一般带有约束，例如，同时在类上标注了@ConditionalOnClass 和@ConditionalOnMissingBean，这样 auto-configuration 只会在 classpath 下存在类并且需要的 Bean 还没有被注入 IoC 时才生效。

Spring Boot 中提供了许多 starter 模块，为开发者提供了许多"一站式"服务。通过在项目添加对应框架的 starter 依赖，可以免去到处寻找依赖包的麻烦。只要加一个依赖项目就可以运行，这就是 starter 的作用。Spring Boot 官方提供的 starter 模块，一般命名规则为"spring-boot-starter -*"，其中"*"代表要使用的应用。

Spring Boot 的 starter 模块包括：

（1）spring-boot-starter，这是 Spring Boot 的核心启动器，包含了自动配置、日志和 YAML。

（2）spring-boot-starter-actuator，帮助监控和管理应用。

（3）spring-boot-starter-amqp，通过 spring-rabbit 来支持 AMQP（Advanced Message Queuing Protocol）。

（4）spring-boot-starter-aop，支持面向方面编程（AOP），包括 spring-aop 和 AspectJ。

（5）spring-boot-starter-artemis，通过 Apache Artemis 支持 JMS 的 API（Java Message Service API）。

（6）spring-boot-starter-batch，支持 Spring Batch，包括 HSQLDB 数据库。

（7）spring-boot-starter-cache，支持 Spring 的 Cache 抽象。

（8）spring-boot-starter-cloud-connectors，支持 Spring Cloud Connectors，简化了在像 Cloud Foundry 或 Heroku 这样的云平台上的连接服务。

（9）spring-boot-starter-data-elasticsearch，支持 ElasticSearch 搜索和分析引擎，包括 spring-data-elasticsearch。

（10）spring-boot-starter-data-gemfire，支持基于 GemFire 的分布式数据存储，包括 spring-data-gemfire。

（11）spring-boot-starter-data-jpa，支持 JPA（Java Persistence API），包括 spring-data-jpa、spring-orm 和 Hibernate。

（12）spring-boot-starter-data-mongodb，支持 MongoDB 数据库，底层使用 MongoDB 驱动操作 MongoDB 数据库，包括 spring-data-mongodb。

（13）spring-boot-starter-data-rest，通过 spring-data-rest-webmvc，支持通过 REST 暴露 Spring Data 数据仓库。

（14）spring-boot-starter-data-solr，支持 Apache Solr 搜索平台，包括 spring-data-solr。

（15）spring-boot-starter-freemarker，支持 FreeMarker 模板引擎。

（16）spring-boot-starter-groovy-templates，支持 Groovy 模板引擎。

（17）spring-boot-starter-hateoas，通过 spring-hateoas 支持基于 HATEOAS 的 RESTful Web 服务。

（18）spring-boot-starter-hornetq，通过 HornetQ 支持 JMS。

（19）spring-boot-starter-integration，支持通用的 spring-integration 模块。

（20）spring-boot-starter-jdbc，支持用 JDBC 访问数据库。

（21）spring-boot-starter-jersey，支持 Jersey RESTful Web 服务框架。

（22）spring-boot-starter-jta-atomikos，通过 Atomikos 支持 JTA 分布式事务处理。

（23）spring-boot-starter-jta-bitronix，通过 Bitronix 支持 JTA 分布式事务处理。

（24）spring-boot-starter-mail，支持 javax.mail。

（25）spring-boot-starter-mobile，支持 spring-mobile。

（26）spring-boot-starter-mustache，支持 Mustache 模板引擎。

（27）spring-boot-starter-redis，支持键值存储数据库 Redis，包括 spring-redis。

（28）spring-boot-starter-security，支持 spring-security。

（29）spring-boot-starter-social-facebook，支持 spring-social-facebook。

（30）spring-boot-starter-social-linkedin，支持 spring-social-linkedin。

（31）spring-boot-starter-social-twitter，支持 spring-social-twitter。

（32）spring-boot-starter-test，支持常规的测试依赖，包括 JUnit、Hamcrest、Mockito 以及 spring-test 模块。

（33）spring-boot-starter-thymeleaf，支持 Thymeleaf 模板引擎，包括与 Spring 的集成。

（34）spring-boot-starter-velocity，支持 Velocity 模板引擎。

（35）spring-boot-starter-web，支持全栈式 Web 开发，包括 Tomcat 和 spring-webmvc。

（36）spring-boot-starter-websocket，支持 WebSocket 开发。

（37）spring-boot-starter-ws，支持 Spring Web Services。

（38）spring-boot-starter-actuator，增加了面向产品上线相关的功能，如测量和监控。

（39）spring-boot-starter-remote-shell，增加了远程 ssh shell 的支持。

（40）spring-boot-starter-jetty，引入了 Jetty HTTP 引擎（可用于替换 Tomcat）。

（41）spring-boot-starter-log4j，支持 Log4J 日志框架。

（42）spring-boot-starter-logging，引入了 Spring Boot 默认的日志框架 Logback。

（43）spring-boot-starter-tomcat，引入了 Spring Boot 默认的 HTTP 引擎 Tomcat。

（44）spring-boot-starter-undertow，引入了 Undertow HTTP 引擎（可用于替换 Tomcat）。

Spring Boot 中 spring-boot-actuator 模块提供了许多附加功能，可以实现在应用程序部署到生产环境后对应用程序进行监控和管理。Spring Boot 提供了 shell 等方式来管理和监控应用程序。另外，审计、监控和性能指标的收集可以自动应用到应用程序上。

Spring Boot 中 spring-boot-cli 模块支持命令行工具 Spring Boot CLI，可以用来快速搭建一个 Spring 原型应用，并且可以运行 Groovy 脚本。

Spring Boot 中 spring-boot-loader 模块允许通过使用 java -jar archive.jar 命令运行包含嵌套依赖的 Jar、War 文件，Spring Boot Loader 提供了三类启动器（Jar、War、Properties Launcher），这些类启动器都可用来加载嵌套在 Jar 里面的资源(如 class 文件、配置文件等)。

Spring Boot 中 tools 模块提供了 Spring Boot 开发者的常用工具集。如 spring-boot-gradle-plugin、spring-boot-maven-plugin 就在这个模块里面。

1.3　Spring Boot 的工作机制

1.3.1　Spring Boot 应用启动入口类的分析

以一个最简单的 Spring Boot 应用为例，程序启动的入口类代码如例 1-1 所示。

【例 1-1】 程序启动的入口类代码示例。

```
@SpringBootApplication
public class Example {
    @RequestMapping("/")
    String home() {
        return "Hello World!";
    }
    public static void main(String[] args) throws Exception {
        SpringApplication.run(Example.class, args);
    }
}
```

分析代码可以发现，注解在代码中占有重要的位置。其中，注解@SpringBootApplication 用于启动 Spring Boot；注解@SpringBootApplication 和@Configuration、@EnableAutoConfiguration、@ComponentScan 等三个注解完全等价（即前面注解是由后面三个注解合并而成）。@RequestMapping 注解用于进行映射关系的设置。

由于注解在代码中占据重要的位置，理解并掌握 Spring Boot 注解的用法是 Spring Boot

学习者必须要迈过去的一道坎。正是由于这个原因，本书在第 3 章专门介绍 Spring Boot 注解相关知识。

1.3.2 Spring Boot 2 的幕后工作

以一个 Spring MVC 应用为例，应用的步骤一般包括：

（1）用户向服务器发送请求，请求被 Spring 前端控制器 DispatcherServlet 捕获；初始化 Spring MVC 的 DispatcherServlet。

（2）搭建转码过滤器，保证客户端的请求正确转码。

（3）视图解析。

（4）配置静态资源（如 CSS 文件）。

（5）配置所支持的地域及资源包（ResourceBundle）。

（6）配置 multipart 解析器，保证文件上传正常工作。

（7）启动内嵌的服务器（Tomcat 或 Jetty 等）。

（8）建立错误反馈页面。

Spring Boot 为 Web 开发者实现了 Spring MVC 应用的所有相关事情。因为这些配置都与应用相关，所以应用 Spring Boot 开发时，可以无限制地组合这些配置。一定程度上，Spring Boot 是带有一定倾向性的 Spring 项目配置器；它基于约定并在项目中默认遵循这些约定。

1.3.3 SpringApplication 的执行流程

SpringApplication 是 Spring Boot 启动的完整解决方案，在没有特殊需求的情况下，SpringApplication 提供的默认执行流程就可以满足 Spring Boot 的需要了。一般来讲，SpringApplication 的执行流程为：

（1）SpringApplication 实例化和初始化。

（2）执行 run()方法。方法开始执行时，首先遍历执行所有通过 SpringFactoriesLoader 可以查找到并加载的 SpringApplicationRunListener，调用它们的 started()方法，告诉这些 SpringApplicationRunListener 即将开始执行 Spring Boot 应用。

（3）配置 Spring Boot 要用到的环境，并在配置好环境后通过遍历调用所有 SpringApplicationRunListener 的 environmentPrepared()的方法，通知 Spring Boot 应用要用的环境已经准备好。

（4）启动并打印 Banner。

（5）根据用户是否明确设置了 applicationContextClass 类型以及初始化阶段的推断结果，决定该为当前 Spring Boot 应用创建什么类型的应用上下文环境，并在配置好后通过遍历调用所有 SpringApplicationRunListener 的 contextPrepared()方法通知 Spring Boot 应用要用的上下文环境已经准备好了。将之前获取的所有配置以及其他形式的 IoC 容器配置加载到已经准备完毕的 ApplicationContext。遍历调用所有 SpringApplicationRunListener 的 contextLoaded()方法。调用 ApplicationContext 的 refresh()方法，完成 IoC 容器可用的最后一道工序。

（6）查看是否有 CommandLineRunner，如果有则遍历它们。

（7）遍历执行 SpringApplicationRunListener 的 finished()方法完成所有功能。如果整个过程出现异常，则在调用所有 SpringApplicationRunListener 的 finished()方法的同时会将异常信息一并传入处理。

1.3.4　Spring Boot 应用启动时控制台输出信息

以 Web 应用为例，Spring Boot 应用启动时控制台输出信息一般包括：

（1）启动 App。
（2）查找 active profile，若无则设为 default。
（3）刷新上下文。
（4）初始化服务器（如 Tomcat）、启动 Tomcat 服务，启动 Servlet。
（5）Spring 内嵌的 WebApplicationContext 初始化。
（6）映射 servlet 和 filter。
（7）查找@ControllerAdvice。
（8）路径映射。
（9）Tomcat 启动完毕。
（10）App 启动耗费的时间。

习题 1

简答题

1. 简述 Spring 的生态圈。
2. 简述 Spring Boot 的特点。
3. 简述 Spring Boot 的 starter 模块组成。
4. 简述 SpringApplication 的执行流程。
5. 以 Web 应用为例，简述 Spring Boot 应用启动时控制台输出信息的组成。

第 2 章

Spring Boot 开发起步

本章主要介绍如何配置 Spring Boot 的开发环境、如何用两款常见开发工具创建项目、如何用两款常见开发工具实现 Hello World 的 Web 应用、以 Hello World 应用为例说明项目属性配置、Spring Boot 开发的一般步骤等内容。

2.1 配置开发环境

Pivotal 团队设计 Spring Boot 的目的是简化 Spring 应用的搭建和开发过程。Spring Boot 使用特定的方式进行配置，从而使开发人员不再需要定义样板化的配置。通过这种方式，Spring Boot 在蓬勃发展的快速应用开发领域成为领导者。Spring Boot 的特点包括创建独立的 Spring 应用程序、自动配置 Spring 等。

在进行 Spring Boot 开发之前，先要配置好开发环境。配置开发环境，需要先安装 JDK，然后选择安装一款合适的开发工具（如 IntelliJ IDEA）。

2.1.1 安装 JDK

使用 2.0.0 以上版本的 Spring Boot 需要安装 1.8 及以上版本的 JDK，可以从 Java 的官网下载安装包。安装完成后，配置环境 JAVA_HOME。配置好 JAVA_HOME 后，将 %JAVA_HOME%\bin 加入系统的环境变量 path 中。完成配置后，打开 Windows 命令处理程序 CMD，输入命令 java -version，如果见到如图 2-1 所示的版本信息就说明 JDK 安装成功了。

```
命令提示符
Microsoft Windows [版本 10.0.16299.431]
(c) 2017 Microsoft Corporation。保留所有权利。

C:\Users\ws>java -version
java version "1.8.0_144"
Java(TM) SE Runtime Environment (build 1.8.0_144-b01)
Java HotSpot(TM) 64-Bit Server VM (build 25.144-b01, mixed mode)

C:\Users\ws>
```

图 2-1　JDK 安装成功后显示的版本信息

2.1.2　安装 IntelliJ IDEA

可以从 IntelliJ IDEA（以下简称为 IDEA）的官网下载免费的社区版或者旗舰试用版 IDEA，然后进行安装，安装完成后打开 IDEA，将显示如图 2-2 所示的欢迎界面。由于 IDEA 自带有 Maven 和 Gradle 插件，所以不用再安装 Maven 和 Gradle 插件。

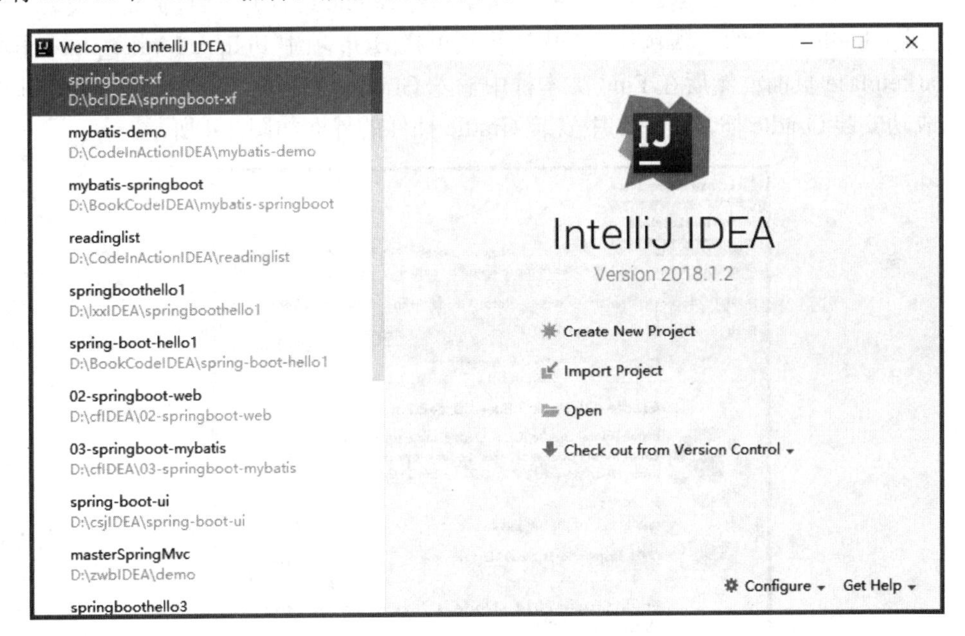

图 2-2　IDEA 启动后的欢迎界面

2.1.3　安装 Spring Tool Suite

Spring Tool Suite（以下简称 STS）是被包装过的 Eclipse，主要用于快速地开发 Spring 项目。利用 STS 开发者不再需要编辑烦琐的 XML 配置文件，而是由工具自动生成。STS 有两种安装方式：一种是在 Eclipse 中安装 STS 插件；另一种是从官网（https://spring.io/tools/sts/all）上直接下载、安装 STS。还可以从官网上直接下载解压缩后即可使用的 STS。STS 启动后的界面如图 2-3 所示。

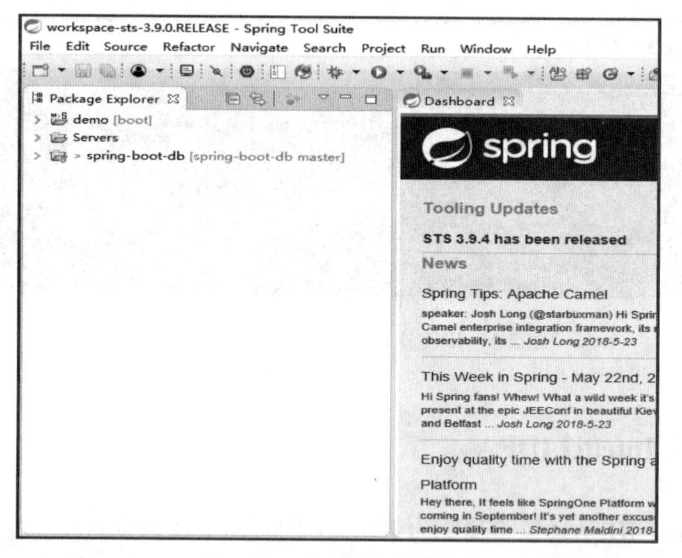

图 2-3 STS 启动后的界面

由于 STS 只是自带有 Maven 插件，所以需要使用 Gradle 进行开发时，需要再安装 Gradle 插件。安装 Gradle 插件的步骤包括：首先选择菜单栏 Help 项中 Eclipse Marketplace 子项后进入 Marketplace 页面；然后在 Find 文本框中输入 Gradle 进行搜索；最后单击 Install 按钮就可以成功安装 Gradle 插件。STS 中安装 Gradle 插件的界面如图 2-4 所示。

图 2-4 STS 中安装 Gradle 插件的界面

2.2　创建项目

视频讲解

可以用不同的工具创建 Spring Boot 项目，每种工具也可以采用不同的创建项目方法。本节介绍两种常见开发工具创建项目的方法。

2.2.1　利用 IDEA 创建项目

先在如图 2-2 所示的欢迎界面中选择 Create New Projcet 链接进入项目创建界面，并选择 Spring Initializr 类型的项目，如图 2-5 所示。

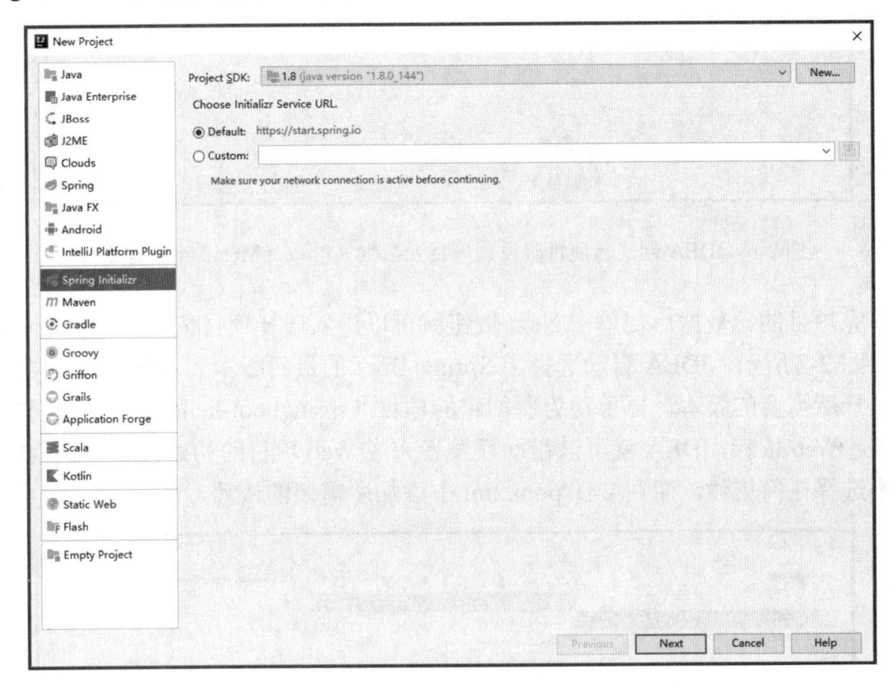

图 2-5　IDEA 中创建新项目时选择 Spring Initializr 类型项目的界面

接着，单击 Next 按钮跳转到项目信息的配置界面，IDEA 创建新项目时要根据项目情况设置项目的元数据（Project Matadata），设置项目元数据的界面如图 2-6 所示。

如图 2-6 所示，在所创建项目 Group 文本框中输入 com.bookcode，在 Artifact 文本框中输入 springboot-helloworlds，在所创建项目的管理工具类型 Type 中选择 Maven Project。由于目前 Maven 的参考资料比 Gradle 的参考资料多且更容易获得，本书示例使用 Maven 进行项目管理。开发语言 Language 选择 Java；打包方式 Packaging 选择 Jar；开发工具 Java 的版本 Java Version 选择 8（也称为 1.8）；所创建项目的版本 Version 保留自动生成的 0.0.1-SNAPSHOT；项目名称 Name 保留自动生成的 springboot-helloworlds；项目描述 Description 可以修改为 Book Code for Spring Boot；所创建项目默认的包名 Package 可以修改为 com.bookcode。

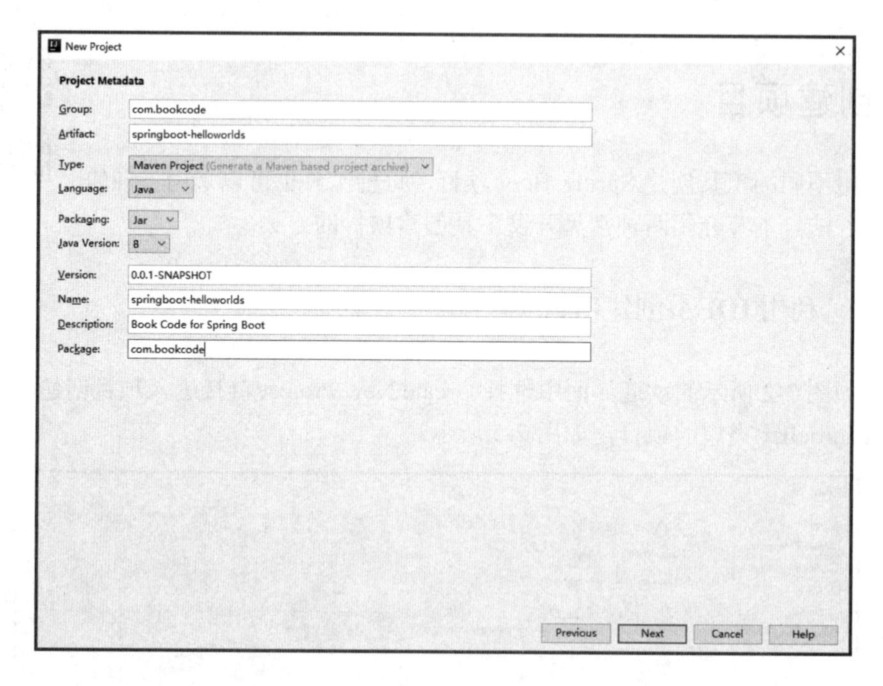

图 2-6　IDEA 创建新项目时设置项目元数据（Project Metadata）的界面

填写完项目的元数据后，单击 Next 按钮就可以进入选择项目依赖（Dependencies）的界面。如图 2-7 所示，IDEA 自动选择了 Spring Boot 的最新版本（本例中 2.0.2 版），也可以手动选择所需要的版本，再手动为所创建的项目（springboot-helloworlds）选择 Web 依赖。选择完 Web 依赖，IDEA 就可以帮助开发者完成 Web 项目的初始化工作。创建项目时，也可以不选择任何依赖，而在文件 pom.xml 中添加所需要的依赖。

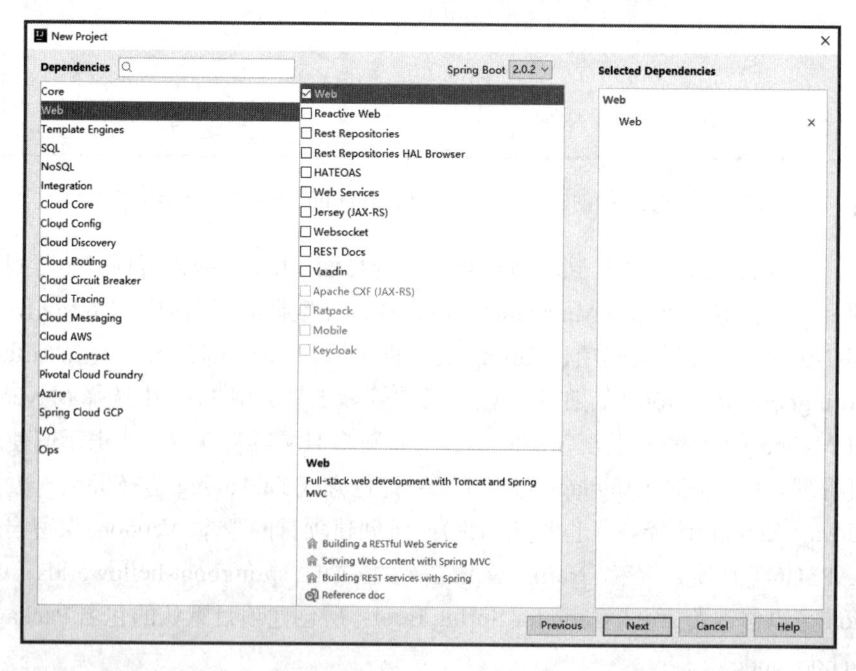

图 2-7　IDEA 创建新项目时选择依赖（Dependencies）的界面

单击 Next 按钮后，进入项目名称（Project Name）和项目位置（Project Location）的显示页面，可以直接保留由图 2-6 生成的项目名称、位置默认值；也可以根据需要直接修改项目名称和项目位置。然后单击 Finish 按钮，就可以进入项目界面。由于所创建的项目管理类型为 Maven Project，所以项目中 pom.xml 文件是一个关键文件，其代码如例 2-1 所示。例 2-1 代码中加粗部分代码和在图 2-6 和图 2-7 中输入、选择的项目元数据和依赖对应，其他代码是 IDEA 自动生成的辅助内容。其中，<parent>和</parent>之间的内容表示父依赖，是一般项目都要用到的基础内容，包含了项目中用到的 Spring Boot 的版本信息。<properties>和</properties>之间的内容表示了项目中所用到的 Java 版本信息和编码格式。<dependencies>和</dependencies>之间的信息是 Maven 的重点内容，包含了项目中所用到的依赖信息，其中<artifactId>spring-boot-starter-test</artifactId>表示要用到测试依赖，测试依赖在创建项目时自动增加，无须再手动添加；<artifactId>spring-boot-starter-web</artifactId>表示要用到 Web 依赖。<build>和</build>之间的内容表示编译运行时要用到的相关插件。关于 Maven 依赖的更多情况，将在后面的例子中逐步加以介绍。在例 2-1 的基础上，就可以进行基于 Spring Boot 的 Web 项目开发了。

【例 2-1】 pom.xml 文件代码示例。

```
<?xml version="1.0" encoding="UTF-8"?>
<project xmlns="http://maven.apache.org/POM/4.0.0"
xmlns:xsi="http://www.w3.org/2001/XMLSchema-instance"
    xsi:schemaLocation="http://maven.apache.org/POM/4.0.0
                        http://maven.apache.org/xsd/maven-4.0.0.xsd">
    <modelVersion>4.0.0</modelVersion>
    <groupId>com.bookcode</groupId>
    <artifactId>springboot-helloworlds</artifactId>
    <version>0.0.1-SNAPSHOT</version>
    <packaging>jar</packaging>
    <name>springboot-helloworlds</name>
    <description>Book Code for Spring Boot</description>
    <parent>
        <groupId>org.springframework.boot</groupId>
        <artifactId>spring-boot-starter-parent</artifactId>
        <version>2.0.2.RELEASE</version>
        <relativePath/> <!-- lookup parent from repository -->
    </parent>
    <properties>
        <project.build.sourceEncoding>UTF-8</project.build.sourceEncoding>
        <project.reporting.outputEncoding>UTF-8</project.reporting.
        outputEncoding>
```

```
        <java.version>1.8</java.version>
    <!--上面加粗内容和图 2-6 中设置的项目元数据对应-->
    </properties>
    <dependencies>
        <!--下面加粗内容和图 2-7 中选择的 Web 依赖对应-->
        <dependency>
            <groupId>org.springframework.boot</groupId>
            <artifactId>spring-boot-starter-web</artifactId>
        </dependency>
        <dependency>
            <groupId>org.springframework.boot</groupId>
            <artifactId>spring-boot-starter-test</artifactId>
            <scope>test</scope>
        </dependency>
    </dependencies>
    <build>
        <plugins>
            <plugin>
                <groupId>org.springframework.boot</groupId>
                <artifactId>spring-boot-maven-plugin</artifactId>
            </plugin>
        </plugins>
    </build>
</project>
```

至此，完成项目的创建工作。在此基础上就可以进行 Spring Boot 的 Web 项目开发了。为了内容的简洁，在本书以后章节的示例和案例中将不再介绍项目的创建过程。假如不太清楚 Spring Boot 项目创建过程，请先熟悉本节内容；或者直接参考源代码中每个项目中 pom.xml 文件的内容。

2.2.2 利用 STS 创建项目

启动 STS，选择菜单项 New 中 Spring Starter Project 子项后可以进入到项目元数据设置界面。STS 设置项目元数据的方法与 IDEA 相似，如图 2-8 所示。单击 Next 按钮进入到选择项目依赖的界面，如图 2-9 所示。选择 Web 依赖，默认情况下 STS 也会自动选择当时最新版 Spring Boot。单击 Finsh 按钮，进入到项目界面。项目中 pom.xml 文件的代码与例 2-1 的代码完全相同。在此基础上，就可以利用 STS 进行基于 Spring Boot 的 Web 项目开发了。

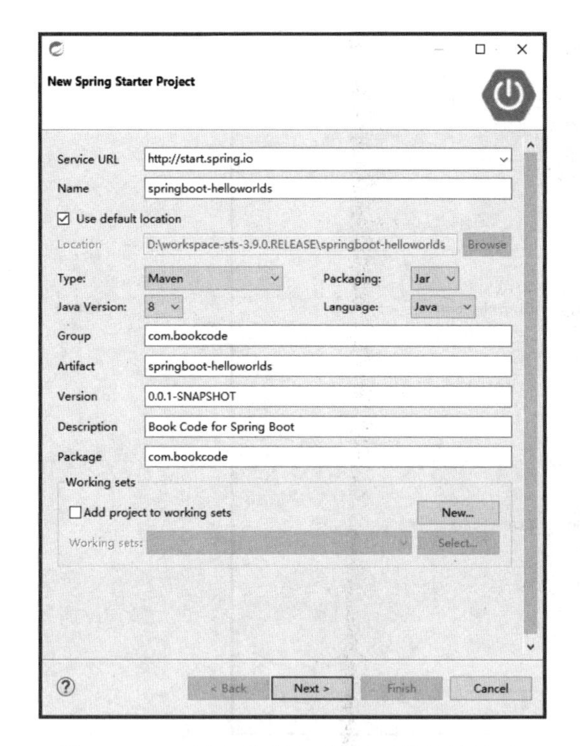

图 2-8　STS 创建新项目时设置元数据的界面

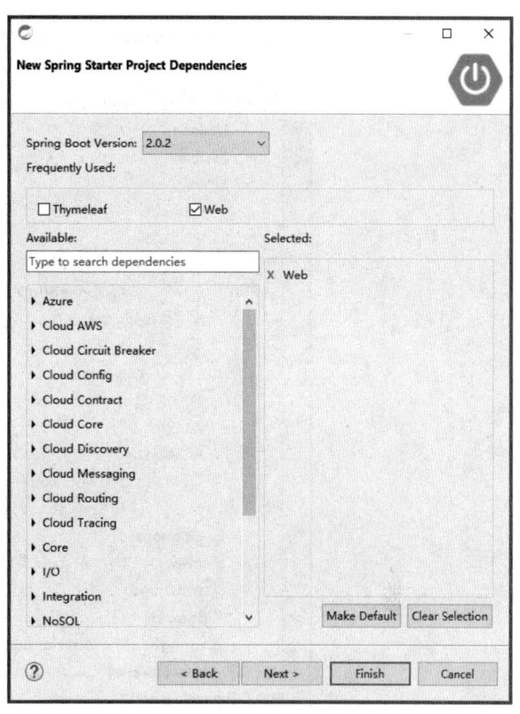

图 2-9　STS 创建新项目时选择依赖的界面

2.3　实现 Hello World 的 Web 应用

视频讲解

与创建项目一样，也可以用不同的工具来开发、实现 Spring Boot 项目。本节介绍如何利用 IDEA、STS 两种工具实现 Hello World 的 Web 应用。

2.3.1　用 IDEA 实现 Hello World 的 Web 应用

IDEA 创建完项目之后，项目中目录和文件的构成情况如图 2-10 所示。Spring Boot 项目中的目录、文件可以分为三大部分。其中，src/main/java 目录下包括主程序入口类 SpringbootHelloworldsApplication，可以运行该类来启动程序；开发时需要在此目录下添加所需的接口、类等文件。src/main/resources 是资源目录，该目录用来存放应用的一些配置信息，如配置服务器端口、数据源的配置文件 application.properties。由于开发的是 Web 应用，因此在 src/main/resources 下产生了 static 子目录与 templates 子目录，static 子目录主要用于存放静态资源，如图片、CSS、JavaScript 等文件；templates 子目录主要用于存放 Web 页面动态视图文件。src/test/java 是单元测试目录，自动生成的测试文件 SpringbootHelloworldsApplicationTests 位于该目录下，用该测试文件可以测试 Spring Boot 应用。另外，pom.xml 文件是项目管理（特别是管理项目依赖）的重要文件。

在自动生成的目录和文件的基础上，在 com.bookcode 包下新建 controller 子包，然后在 controller 子包中创建类 HelloWorldController，修改后的代码如例 2-2 所示。

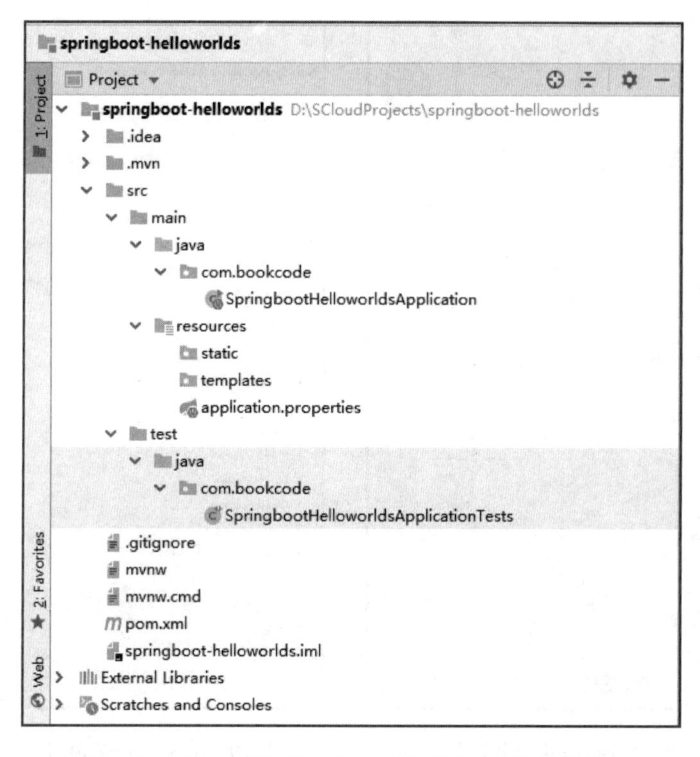

图 2-10　IDEA 创建项目后项目的目录和文件构成情况

【例 2-2】 控制器类 HelloWorldController 修改后的代码示例。

```
package com.bookcode.controller;
import org.springframework.web.bind.annotation.RequestMapping;
import org.springframework.web.bind.annotation.RestController;
@RestController                                   //返回的默认结果为字符串
public class HelloWorldController {
    @RequestMapping("/hello")                     //映射信息，往往是 URL 的组成部分
    public String hello(){
        return "Hello World!";
    }
}
```

接着运行入口类 SpringbootHelloworldsApplication，成功启动自带的内置 Tomcat。在浏览器中输入 localhost:8080/hello 后，浏览器中显示的 Web 应用运行结果如图 2-11 所示。

图 2-11　IDEA 实现 Hello World 的 Web 应用运行结果

2.3.2　用 STS 实现 Hello World 的 Web 应用

STS 创建完项目之后，项目的目录和文件结构如图 2-12 所示，和 IDEA 实现的项目情况基本相同。在自动生成文件的基础上，在 com.bookcode 包下新建 controller 子包，然后在 controller 子包中创建一个 HelloWorldController 类，代码如例 2-2 所示。

接着运行程序，成功启动内置 Tomcat，再在浏览器中输入 localhost:8080/hello，浏览器中的结果如图 2-11 所示。

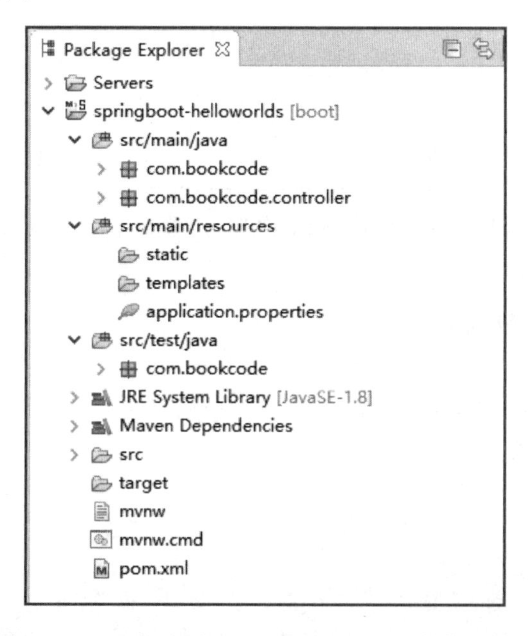

图 2-12　STS 创建项目后项目的目录和文件构成情况

由于用 STS 和 IDEA 开发 Spring Boot 的方法类似，考虑到 IDEA 的功能更丰富，本书选用 IDEA 作为开发工具。

2.4　以 Hello World 应用为例说明项目属性配置

视频讲解

在实现 Hello World 应用的基础上，可以基于项目属性配置实现对 Hello World 应用的扩展。在 Spring Boot 中主要通过 application.properties 文件、application.yml 文件实现对属性的配置。这两种文件的格式不同，但内容对应、作用相同。配置文件的默认执行顺序依次是：项目根目录下 config 子目录、项目根目录、项目 classpath 子目录下的 config 子目录、项目 classpath 子目录。

2.4.1　配置项目内置属性

可以修改 application.properties 文件，配置项目内置属性，代码如例 2-3 所示。

本书后面章节中修改文件或类的代码均指修改后的代码。

【例 2-3】 配置项目端口、路径等内置属性的代码示例。

```
#配置项目内置属性，修改端口
server.port=8888
server.servlet.context-path=/website
```

运行程序后，在浏览器中输入 localhost:8888/website/hello，结果如图 2-13 所示。对比例 2-2、例 2-3 中代码和图 2-11、图 2-13 中 URL，可以发现例 2-3 通过配置文件修改了服务器默认的端口和路径。

图 2-13 修改 Web 应用 Hello World 的服务器默认端口和路径配置后的结果

2.4.2 自定义属性设置

可以修改 application.properties 文件来自定义项目属性，代码如例 2-4 所示。

【例 2-4】 自定义项目属性的代码示例。

```
#自定义属性
server.port=8888
server.servlet.context-path=/website
helloWorld=Hello SpringBoot!
mysql.jdbcName=com.mysql.jdbc.Driver
mysql.dbUrl=jdbc:mysql://localhost:3306/mytest
mysql.userName=root
mysql.password=sa
```

再修改类 HelloWorldController，修改后的代码如例 2-5 所示。

【例 2-5】 类 HelloWorldController 修改后的代码示例。

```
package com.bookcode.controller;
import org.springframework.web.bind.annotation.RequestMapping;
import org.springframework.web.bind.annotation.RestController;
import org.springframework.beans.factory.annotation.Value;
@RestController                                    //返回的默认结果为字符串
public class HelloWorldController {
@Value("${helloWorld}")                            //注入属性值
    private String hello;
@Value("${mysql.jdbcName}")
    private String jdbcName;
@Value("${mysql.dbUrl}")
    private String dbUrl;
```

```
@Value("${mysql.userName}")
    private String userName;
@Value("${mysql.password}")
    private String password;
@RequestMapping("/hello")                        //映射信息，往往是 URL 的组成部分
    public String hello(){
        return  hello;
    }
@RequestMapping("/showJdbc")                     //映射信息，往往是 URL 的组成部分
    public String showJdbc(){
        return "mysql.jdbcName:"+jdbcName+"<br/>"
                +"mysql.dbUrl:"+dbUrl+"<br/>"
                +"mysql.userName:"+userName+"<br/>"
                +"mysql.password:"+password+"<br/>";
    }
}
```

接着运行程序，在浏览器中输入 localhost:8888/website/hello，结果如图 2-14 所示。在浏览器中输入 localhost:8888/website/showJdbc，结果如图 2-15 所示。

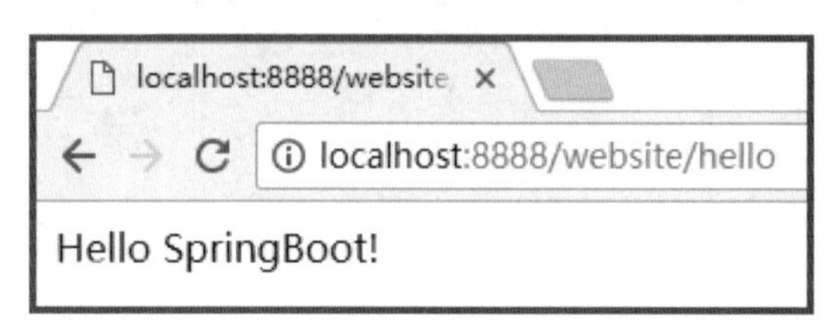

图 2-14　IDEA 中自定义属性后在浏览器中输入 localhost:8888/website/hello 的结果

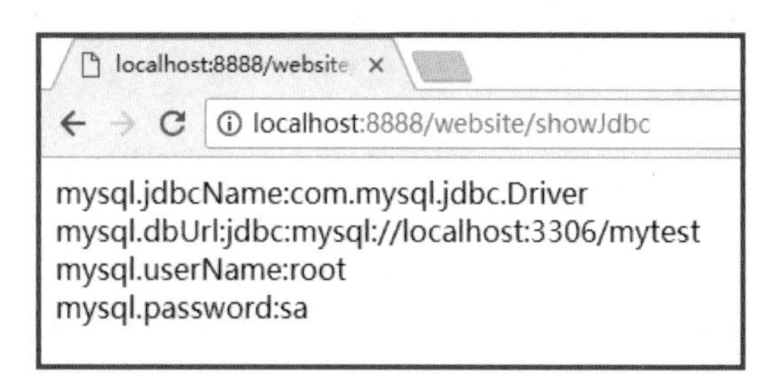

图 2-15　IDEA 中自定义属性后在浏览器中输入 localhost:8888/website/showJdbc 的结果

2.4.3　利用自定义配置类进行属性设置

当要配置的自定义属性比较多时（如例 2-4），可以考虑自定义一个配置类。自定义配置类 MysqlProperties 的代码如例 2-6 所示。

【例 2-6】 自定义配置类 MysqlProperties 的代码示例。

```
//自定义配置类
package com.bookcode.properties;
```

```
import org.springframework.boot.context.properties.ConfigurationProperties;
import org.springframework.stereotype.Component;
@Component                                          //构件
@ConfigurationProperties(prefix="mysql")            //配置属性类
public class MysqlProperties {
   private String jdbcName;
   private String dbUrl;
   private String userName;
   private String password;
   public String getJdbcName() {
      return jdbcName;
   }
   public void setJdbcName(String jdbcName) {
      this.jdbcName = jdbcName;
   }
   public String getDbUrl() {
      return dbUrl;
   }
   public void setDbUrl(String dbUrl) {
      this.dbUrl = dbUrl;
   }
   public String getUserName() {
      return userName;
   }
   public void setUserName(String userName) {
      this.userName = userName;
   }
   public String getPassword() {
      return password;
   }
   public void setPassword(String password) {
      this.password = password;
   }
}
```

然后修改类 HelloWorldController，代码如例 2-7 所示。

【例 2-7】 类 HelloWorldController 的代码示例。

```
package com.bookcode.controller;
import org.springframework.web.bind.annotation.RequestMapping;
import org.springframework.web.bind.annotation.RestController;
import com.bookcode.properties.MysqlProperties;
import org.springframework.beans.factory.annotation.Value;
import javax.annotation.Resource;
```

```
@RestController                              //返回的默认结果为字符串
public class HelloWorldController {
@Value("${helloWorld}")                      //注入属性值
    private String hello;
@Resource                                    //属性值
    private MysqlProperties mysqlProperties;
@RequestMapping("/hello")                    //映射信息，往往是 URL 的组成部分
    public String hello(){
        return  hello;
    }
@RequestMapping("/showJdbc")                 //映射信息，往往是 URL 的组成部分
    public String showJdbc(){
        mysqlProperties.setJdbcName("com.mysql.jdbc.Driver");
        mysqlProperties.setDbUrl("jdbc:mysql://localhost:3306/mytest");
        mysqlProperties.setUserName("root");
        mysqlProperties.setPassword("sa");
    return "mysql.jdbcName:"+mysqlProperties.getJdbcName()+"<br/>"
            +"mysql.dbUrl:"+mysqlProperties.getDbUrl()+"<br/>"
            +"mysql.userName:"+mysqlProperties.getUserName()+"<br/>"
            +"mysql.password:"+mysqlProperties.getPassword()+"<br/>";
    }
}
```

删除配置文件中相关信息，代码如例 2-8 所示（请对照例 2-4 的代码）。

【例 2-8】 自定义属性代码示例。

```
#自定义属性
server.port=8888
server.servlet.context-path=/website
helloWorld=Hello SpringBoot!
```

运行程序后在浏览器中输入 localhost:8888/website/hello，结果如图 2-14 所示。在浏览器中输入 localhost:8888/website/showJdbc，结果如图 2-15 所示。

2.5 Spring Boot 开发的一般步骤

视频讲解

2.5.1 软件生命周期

为了应对软件危机，产生了软件工程学。软件工程是指导计算机软件开发和维护的一门工程学科。采用工程的概念、原理、技术和方法来开发与维护软件；把经过时间考验的管理技术和当前最适用的技术方法结合起来，以经济地开发出高质量的软件并有效地维护它，这就是软件工程。

按照软件工程的理论，软件生命周期由软件定义、软件开发和运行维护（也称为软件维护）3 个时期组成，每个时期又进一步划分成若干阶段。

软件定义时期的任务是：确定软件项目必须完成的总目标；确定工程的可行性；导出实现项目目标应该采用的策略及系统必须完成的功能；估计完成项目需要的资源和成本，并且制订工程进度表。这个时期的工作通常又称为系统分析，由系统分析员负责完成。软件定义时期通常进一步划分成 3 个阶段，即问题定义、可行性研究和需求分析。

软件开发时期具体设计和实现在前一个时期定义的软件，它通常由下述 4 个阶段组成：总体设计、详细设计、编码和单元测试、综合测试。其中前两个阶段又称为系统设计，后两个阶段又称为系统实现。

运行维护时期的主要任务是使软件持久地满足用户的需要。通常运行维护活动包括改正性维护、适应性维护、完善性维护和预防性维护。

由于本书主要是探讨基于 Spring Boot 的应用开发，所以本书介绍的示例和案例主要说明的是如何用 Spring Boot 进行编码实现（简称 Spring Boot 开发）。

2.5.2 Spring Boot 开发步骤

Spring Boot 的开发步骤如下。

Step 1：打开开发工具。

Step 2：创建项目。

Step 3：根据情况判断是否需要添加（补充）项目所需的依赖，如果没有需要补充的依赖则跳过此步骤。

Step 4：创建类和接口（按照实体类、数据访问接口和类、业务接口和类、控制器类等顺序）。

Step 5：根据情况判断是否需要创建视图文件和 CSS 文件等，如果不需要则跳过此步骤。

Step 6：根据情况判断是否需要创建配置文件，如果不需要则跳过此步骤。

Step 7：根据情况判断是否需要图片、语音、视频等文件，如果不需要则跳过此步骤。

Step 8：根据情况判断是否需要下载辅助文件、包和安装工具（如数据库 MySQL），如果不需要则跳过此步骤。由于本书中用到的工具较多且安装使用比较简单，本书对此步骤介绍比较少。

要注意的是，Step 3～Step 8 的 5 个步骤之间的顺序可以互换。完成了 Spring Boot 开发之后，就可以运行程序了。

习题 2

简答题

1. 简述属性配置的不同方法。

2．简述 Spring Boot 的开发步骤。

实验题

1．安装开发工具 IDEA。

2．安装或配置开发工具 STS。

3．用 IDEA 实现 Hello World 的 Web 应用。

4．用 STS 实现 Hello World 的 Web 应用。

5．以 Hello World 应用为例，用不同方式实现项目属性配置。

第 3 章

Spring Boot 的相关注解

对第 1 章中例 1-1 的代码进行分析，可以知道注解（Annotation）在 Spring Boot 开发中占有重要的地位。注解为在代码中添加信息提供了一种形式化的方法，通过注解可以很方便地在代码中某个地方使用被注解的对象。注解常见的作用包括生成文档、跟踪代码依赖性和替代配置文件、在编译时进行格式检查（如@Override 放在方法前）等。

为了帮助读者更好地理解后面章节的示例代码，本章介绍 Spring Boot 注解以及和 Spring Boot 注解密切相关的 Java 注解、Spring 注解等内容。考虑到注解使用频率较高，本章将@RestController、@Bean 等 Spring 注解放到 Spring Boot 注解中加以说明。

3.1 Java 注解

JRE 的库包 java.lang.annotation 中代码包括注解相关的接口、类等内容。接口 java.lang.annotation.Annotation 是所有自定义注解自动继承的接口，不需要定义时指定，类似于所有 Java 类都自动继承的 Object 类。

3.1.1 Java 注解的介绍

注解是一系列元数据，它利用元数据来解释、说明程序代码（即被注解的对象）。但是，注解不是所标注的代码的组成部分。注解对于代码的运行效果没有直接影响。注解的作用包括：

（1）提供信息给编译器，编译器可以利用注解来探测错误和警告信息。

（2）软件工具可以利用注解信息来生成代码、HTML 文档或者做其他相应处理。

（3）运行时的处理，某些注解可以在程序运行时接受代码的提取。

在开始学习注解具体语法之前，可以把注解看成一张标签。与接口和类一样，注解也是一种类型。注解是自 Java 1.5 开始引入的概念，它允许开发者定义自己的注解类型和使用自定义的注解。注解通过关键字@interface 进行定义，代码如例 3-1 所示。

【例 3-1】 注解定义的代码示例。

```
public @interface TestAnnotation {
}
```

例 3-1 中的代码自定义了一个名字为 TestAnnotaion 的注解，该注解自动继承了类 java.lang.annotation.Annotation。创建了自定义注解后，就可以使用自定义的注解。注解的使用方法如例 3-2 所示。

【例 3-2】 使用自定义注解的代码示例。

```
@TestAnnotation
public class Test {
}
```

例 3-2 创建了一个类 Test，并在类定义的上方加上了 TestAnnotation 注解，就意味着用 TestAnnotation 注解了类 Test。

3.1.2　Java 的元注解

要想更好地使用注解，还需要理解元注解。元注解是加到注解上的注解，它的目的是解释、说明其他普通注解。元注解有@Retention、@Documented、@Target、@Inherited、@Repeatable 共 5 种。

元注解@Retention 应用到一个注解时，说明该注解的存活时间。它的取值包括 RetentionPolicy.SOURCE、RetentionPolicy.CLASS、RetentionPolicy.RUNTIME。其中，RetentionPolicy.SOURCE 表明注解只在源码阶段保留，在编译器进行编译时被丢弃。RetentionPolicy.CLASS 表明注解被保留到编译进行的时候，而不会被加载到 JVM 中。RetentionPolicy.RUNTIME 表明注解可以保留到程序运行的时候，它会被加载进入 JVM 中；在程序运行时可以获取到它们。

应用元注解@Retention 的代码示例如例 3-3 所示，在该示例中设定注解 TestAnnotation 可以在程序运行时被获取到。

【例 3-3】 应用元注解@Retention 的代码示例。

```
@Retention(RetentionPolicy.RUNTIME)
public @interface TestAnnotation {
}
```

元注解@Documented 表示注解内容会被 Javadoc 工具提取成文档，文档内容会因为注解内容的不同而不同。

元注解@Target 表示注解用于什么地方，如类型、方法和域等。元注解@Target 的取值包括 ElementType.FIELD、ElementType.METHOD、ElementType.PARAMETER、ElementType.CONSTRUCTOR、ElementType.LOCAL_VARIABLE、ElementType.TYPE、ElementType.ANNOTATION_TYPE、ElementType.PACKAGE。其中，ElementType.FIELD

表示对字段、枚举常量的注解，ElementType.METHOD 表示对方法的注解，ElementType.PARAMETER 表示对方法参数的注解，ElementType.CONSTRUCTOR 表示对构造函数的注解，ElementType.LOCAL_VARIABLE 表示对局部变量的注解，ElementType.ANNOTATION_TYPE 表示对注解类型的注解，ElementType.PACKAGE 表示对包的注解，ElementType.TYPE 表示对接口、类、枚举、注解等任意类型的注解。

被元注解@Inherited 注解过的注解作用于父类后，子类会自动继承父类的注解。应用的示例代码如例 3-4 所示。

【例 3-4】 应用元注解@Inherited 的代码示例。

```
//该示例由 3 部分组成
//定义注解
@Inherited
@Retention(RetentionPolicy.RUNTIME)
@interface Test {}

//在父类中增加注解
@Test
public class A {}

//子类 B 虽然没有明确给出注解信息，但是它会继承父类 A 中的注解@Test
public class B extends A {}
```

元注解@Repeatable 是在 Java 1.8 中引入的注解，其应用的示例代码如例 3-5 所示。Persons 可以被看作是一张总标签，上面贴满了 Person 这种同类型但内容不一样的标签。于是，可以同时给 SuperMan 贴上画家、程序员、产品经理等标签（即加上 3 个注解）。

【例 3-5】 应用元注解@Repeatable 的代码示例。

```
//Persons 是用数组存放注解的容器注解，它里面必须要有一个 value 属性，即数组
@interface Persons {
    Person[]  value();
}
//可以重复、多次应用 Persons 注解
@Repeatable(Persons.class)
@interface Person{
    String role default "";
}
//不同属性的 Person 注解
@Person(role="artist")
@Person(role="coder")
@Person(role="PM")
public class SuperMan{
}
```

注解中可以拥有属性（也叫作成员变量），示例代码如例 3-6 所示。表示自定义的注解 TestAnnotation 注解拥有 id 和 msg 两个属性，返回类型分别为 int 和 String。

【例 3-6】 带属性的注解代码示例。

```
@Target(ElementType.TYPE)
@Retention(RetentionPolicy.RUNTIME)
public @interface TestAnnotation {
    int id();
    String msg();
}
```

使用注解时，应该给属性赋值。属性的赋值方法是在注解后面的括号内以“属性=取值”的形式进行的，多个属性的赋值之间用逗号隔开，示例代码如例 3-7 所示。需要注意的是，注解中属性的类型只能是 8 种基本数据类型和类、接口、注解及它们的数组。

【例 3-7】 给注解属性赋值的应用代码示例。

```
@TestAnnotation(id=3,msg="hello annotation")
public class Test {
}
```

注解中属性可以有默认值，默认值需要用关键字 default 指定，示例代码如例 3-8 所示。注解 TestAnnotation 中属性 id 的默认值被指定为 −1，属性 msg 的默认值被指定为"Hi"。

【例 3-8】 给注解属性指定默认值的代码示例。

```
@Target(ElementType.TYPE)
@Retention(RetentionPolicy.RUNTIME)
public @interface TestAnnotation {
    public int id() default -1;
    public String msg() default "Hi";
}
```

假如指定了属性默认值，可以不用再在注解 TestAnnotation 后面的括号内对属性进行赋值。示例代码如例 3-9 所示。

【例 3-9】 属性有默认值的注解应用代码示例。

```
@TestAnnotation()
public class Test {}
```

如果注解只有一个属性时，应用这个注解时则可以省略属性名而将属性值直接填写到注解后面的括号内，如例 3-10 所示。

【例 3-10】 单一属性注解简单应用的代码示例。

```
@Check("hi")
int a;
```

例 3-11 代码和例 3-10 代码的效果是一样的。

【例 3-11】 单一属性注解完整应用的代码示例。

```
@Check(value="hi")
int a;
```

如果一个注解没有任何属性，应用这个注解时注解后面的括号则可以省略，如例 3-12 所示。

【例 3-12】 无属性注解的应用代码示例。

```
@Perform
public void testMethod(){}
```

3.1.3 Java 预置的基本注解

在 java.lang 包下，Java 预先提供了@Deprecated、@SuppressWarnings、@Override、@SafeVarargs、@FunctionalInterface 共 5 个基本注解。

注解@Deprecated 是用来标记过时的元素。编译器在编译阶段遇到这个注解时会发出提醒警告，告诉开发者正在调用一个过时的元素（如过时的方法、类、成员变量）。

注解@SuppressWarnings 表示阻止警告的意思。使用@Deprecated 注解后，编译器有时会给开发者发出警告提醒；当开发者想忽略警告提醒时，可以通过 @SuppressWarnings 注解达到目的。@SuppressWarnings 的参数有 deprecation（表示忽略使用了过时元素时的警告）、unchecked（表示忽略执行了未检查转换时的警告）、fallthrough（表示忽略 switch 程序块直接通往下一种情况而没有 break 时的警告）、path（表示忽略类路径、源文件路径等路径中有不存在的路径时的警告）、serial（表示忽略可序列化的类缺少 serialVersionUID 定义时的警告）、finally（表示忽略任何 finally 子句不能正常完成时的警告）、all（表示忽略关于所有情况的警告）。

注解@Override 表示子类要重写父类（或接口）的对应方法。如果想重写父类的方法，如 toString()方法，则在方法前面加上@Override，系统可以帮助检查方法的正确性。示例代码如例 3-13 所示。

【例 3-13】 注解@Override 的应用代码示例。

```
//@Override 可以帮助检查方法的正确性
@Override
public String toString(){...} //这是子类方法正确的写法

//下面子类方法是错误的，有@Override，系统可以帮助检查出 tostring 的拼写错误
@Override
public String tostring(){...}

//下面子类方法是错误的，由于没有@Override，系统不会帮助检查出 tostring 的拼写错误
public String tostring(){...}
```

注解@SafeVarargs 被用来标识参数安全类型，它被用来提醒开发者不要用参数做一些不安全的操作，它的存在会阻止编译器产生 unchecked 这样的警告。它是在 Java 1.7 中引入的注解。

注解@FunctionalInterface 被用来指定某个接口必须是函数式接口，否则就会编译出错。@FunctionalInterface 是 Java 1.8 引入的新特性。如果接口中有且只有一个抽象方法（可以

包含多个默认方法或多个 static 方法），该接口称为函数式接口。

3.2　Spring 注解及注解注入

Spring 容器通过把 Java 类注册成 Bean 的方式来管理 Java 类。把 Java 类变成 Bean 有两种方式：一种方式通过 XML 配置把 Java 类注册成 Bean；另一种方式通过注解的方法将 Java 类注册成 Bean。注解方式下，只需要在 Java 类前边加上注解，Spring 扫描到注解后就会将被注解的类自动注册成 Bean。利用不同的注解可以将 Java 类注册成不同的 Bean。相对于 XML 配置，注解方法更加方便、快捷。于是，越来越多的工具都支持用注解进行配置而放弃 XML 配置。

3.2.1　Spring 基础注解

可以将注解@Component 放在类的前面，该类被标注成 Spring 的一个普通 Bean。

注解@Controller 标注一个控制器组件类，它被用来实现自动检测类路径下的组件并将组件自动注册成 Bean。例如，使用@Controller 注解标注 Java 类 UserAction 之后，就表示要把类 UserAction 标注成 Bean 后交给 Spring 容器管理。如果不指定 Bean 的名字，按照约定该 Bean 会被命名为 userAction。也可以在注解后面的括号内指定 Bean 的名字，例如采用@Controller(value="UA")或者@Controller("UA")的方法来指定 Bean 的名字为 UA。

注解@Service 标注一个业务逻辑组件类。例如，类 Action 需要使用 UserServiceImpl 实例时，可以用@Service("userService")注解告诉 Spring 创建好一个 UserServiceImpl 实例（名字为 userService）。Spring 创建好 userService 之后可以将其注入给 Action，Action 就可以使用该 UserServiceImpl 实例了。

注解@Repository 标注一个 DAO 组件类。可以用@Repository(value="userDao")注解告诉 Spring 创建一个 UserDao 实例（名字为 userDao）。当 Service 需要使用 userDao 时，可以用@Resource(name = "userDao")注解告诉 Spring 把创建好的 userDao 注入给 Service。

3.2.2　Spring 常见注解

注解@Autowired 被用来实现自动装配，@Autowired 可以被用来标注成员变量、方法、构造函数等对象。虽然@Autowired 的标注对象不同，但是都会在 Spring 初始化 Bean 时进行自动装配。使用@Autowired 可以使 Spring 容器自动搜索符合要求的 Bean，并将其作为参数注入。

@Autowired 是根据类型进行自动装配的，示例代码如例 3-14 所示。如果 Spring 上下文中存在多个同类型的 Bean 时（如有两个类都实现了 EmployeeService 接口），Spring 不知道应该绑定哪个实现类，就会抛出 BeanCreationException 异常。如果 Spring 上下文中不存在某个类型的 Bean 时，也会抛出 BeanCreationException 异常。

【例 3-14】 注解@Autowired 的应用代码示例。

```
//接口 EmployeeService 声明
public interface EmployeeService {
    public EmployeeDto getEmployeeById(Long id);
}

//两个实现类 EmployeeServiceImpl 和 EmployeeServiceImpl1
@Service("service")
public class EmployeeServiceImpl implements EmployeeService {
    public EmployeeDto getEmployeeById(Long id) {
        return new EmployeeDto();
    }
}
@Service("service1")
public class EmployeeServiceImpl1 implements EmployeeService {
    public EmployeeDto getEmployeeById(Long id) {
        return new EmployeeDto();
    }
}

//调用接口实现类
@Controller
@RequestMapping("/emplayee.do")
public class EmployeeInfoControl {

    @Autowired          //此注解处会出错，因为有两个实现类，而不知道绑定哪一个
    EmployeeService employeeService;

    @RequestMapping(params = "method=showEmplayeeInfo")
    public void showEmplayeeInfo(HttpServletRequest request, HttpServletResponse
    response, EmployeeDto dto) {
        …//代码省略
    }
}
```

可以使用@Qualifier 配合@Autowired 来解决异常 BeanCreationException，示例代码如例 3-15 所示。

【例 3-15】 注解@Qualifier 和@Autowired 配合使用的代码示例。

```
//接口 EmployeeService 声明与例 3-14 代码相同
//接口的两个实现类 EmployeeServiceImpl 和 EmployeeServiceImpl1 与例 3-14 代码相同

//接口实现类的调用
@Controller
@RequestMapping("/emplayee.do")
```

```
public class EmployeeInfoControl {

    @Autowired
    @Qualifier("service")                  //新增加语句，指定调用第一个接口实现类
    EmployeeService employeeService;

    @RequestMapping(params = "method=showEmplayeeInfo")
    public void showEmplayeeInfo(HttpServletRequest request, HttpServletResponse
    response, EmployeeDto dto) {
        …//代码省略
    }
}
```

注解@Resource 可用于标注一个对象的 SET 方法。注解@Resource 的作用相当于@Autowired，只不过@Autowired 按类型自动注入，而@Resource 默认按名字自动注入。JSR-250 标准推荐使用通用注解@Resource 来代替 Spring 专有的@Autowired 注解。

JSR（Java Specification Requests，Java 规范提案）是指向 JCP（Java Community Process）提出新增一个标准化技术规范的正式请求。任何人都可以提交 JSR，以向 Java 平台增添新的 API 和服务。JSR 已成为 Java 界的一个重要标准。从 Spring 2.5 开始，Spring 框架的核心支持 JSR-250 标准中注解@Resource、注解@PostConstruct 和注解@PreDestroy。

@Resource 有两个属性比较重要，分别是名字 name 和属性 type。Spring 将@Resource 注解 name 属性解析为 Bean 的名字，而将 type 属性解析为 Bean 的类型。如果使用 name 属性，则使用按名字自动注入的注入策略，而使用 type 属性时则使用按类型自动注入的注入策略。如果既不指定 name 属性也不指定 type 属性，则通过反射机制使用按名字自动注入的注入策略。@Resource 使用按名字自动注入的注入策略时，与使用@Qualifier 明确指定 Bean 的名称进行注入作用相同。在众多相同的 Bean 中，优先使用@Primary 注解的 Bean。这和@Qualifier 有点区别，@Qualifier 指的是使用哪个 Bean 进行注入。

注解@PostConstruct 和注解@PreDestroy 都是 JSR-250 标准的注解。标注了 @PreDestroy 的方法将在类销毁之前调用，@PostConstruct 注解过的方法将在类实例化后调用。

从 Spring 3.0 开始，Spring 开始支持 JSR-330 标准的注解。JSR-330 中，@Inject 和 Spring 中的@Autowired 的职责相同，@Named 和 Spring 中的@Component 的职责类似。

可以用注解@Scope 来定义 Bean 的作用范围（称为作用域），也可以通过在 XML 文件中设置 Bean 的 scope 属性值来实现这一目的。@Scope 注解的值可以是 singleton、prototype、request、session、global session 等作用域，默认是单例模式，即 scope="singleton"。其中，单例模式 singleton 表示全局有且仅有一个实例；原型模式 prototype 表示每次获取 Bean 时都会有一个新的实例；request 表示针对每一次 HTTP 请求都会产生一个新的 Bean，而且该 Bean 仅在当前 HTTP 请求内有效；session 表示针对每一次 HTTP 请求都会产生一个新的 Bean，而且该 Bean 仅在当前 HTTP session 内有效；global session 类似于标准的 HTTP session，不过它仅仅在基于 Portlet 的 Web 应用中才有意义。Portlet 的请求处理分为 action 阶段和 render 阶段。在一个请求中，action 阶段只执行一次，但是 render 阶段可能由于用

户的浏览器操作而被执行多次。Portlet 规范定义了 global session 的概念，它被所有构成某个 Portlet Web 应用的各种不同 Portlet 所共享。在 global session 中定义的 Bean 被限定于 Portlet 全局会话（global session）的生命周期范围内。如果在 Web 中使用 global session 来标识 Bean，那么 Web 会自动当成 session 类型来使用。

JSR-330 默认的作用域类似 Spring 的 prototype，JSR-330 标准中的 Bean 在 Spring 中默认也是单例的。如果要使用非单例的作用域，开发者应该使用 Spring 的@Scope 注解。JSR-330 也提供了一个@Scope 注解，然而，这个注解仅仅在用来创建自定义的作用域时才能使用。

注解@RequestMapping 为类或方法指定一个映射路径，可以通过指定的路径来访问对应的类或方法，应用示例如例 3-16 所示。其中，userid 值通过@PathVariable 注解方法进行绑定。注解@PathVariable 主要用来获取单一的 URI 参数，如果想通过 URI 传输一些复杂的参数，则要考虑使用注解@MatrixVariable。@MatrixVariable 的矩阵变量可以出现在 URI 中任何地方，变量之间用分号（;）分隔。@MatrixVariable 默认是不启用的，启用它时需要将 enable-matrix-variables 设置为 true。

【例 3-16】 注解@RequestMapping 的应用代码示例。

```
@RequestMapping(value="/getName/{userid}", method = RequestMethod.GET)
public void login(@PathVariable String userid, Model model){
}
```

注解@RequestParam 将请求中带的值赋给被注解的方法参数。如例 3-17 所示，把请求中的值 username 赋给方法中 username 这个参数。属性 required 代表参数是否必须赋值，默认为 true；当不能确定请求中是否有值可以赋给参数时，就必须把属性 required 设置为 false。

【例 3-17】 注解@RequestParam 的应用代码示例。

```
public void login(@RequestParam(value="username" required="true")String
username){
}
```

不管是 HTTP 请求还是 HTTP 响应都是通过报文传输的，而报文都有头和正文。头包含服务端或者客户端的一些信息，用来表明身份、提供验证或限制等。正文是要传递的内容。注解@RequestBody 把请求报文中的正文自动转换成绑定给方法参数的变量字符串。响应请求时，@ResponseBody 将内容或 Java 对象转换成响应报文的正文返回。当返回的数据不是 HTML 标记页面（视图）而是其他某种格式数据（如 JSON、XML 等）时，才使用@RequestBody 注解。

注解@Param 表示对参数的解释，一般写在注释里面。

注解@JoinTable 表示 Java 类和数据库表的映射关系，也可以标识列的映射、主键的映射等。

注解@Transactional 是 Spring 事务管理的注解。被@Transactional 注解的方法或类自动被注册成事务，接受 Spring 容器的管理。

关键字 Synchronized 表示实现 Java 同步机制，它不能和注解@Transactional 同时使用。

注解@ModelAttribute 声明在属性上，表示该属性的值来源于 model 里 queryBean，并被保存到 model 里。注解@ModelAttribute 声明在方法上，表示该方法的返回值被保存到 model 里。

注解@Cacheable 表明一个方法的返回值应该被缓存，注解@CacheFlush 声明一个方法是清空缓存的触发器，这两个注解要配合缓存处理器使用。

Spring 允许指定 ModelMap 中哪些属性需要转存到会话中，以便下一个请求还能访问到这些属性。注解@SessionAttributes 只能标注类，而不能标注方法。

如果希望某个属性编辑器仅作用于特定的控制器 Controller，可以在 Controller 中定义一个被注解@InitBinder 标注的方法，可以在该方法中向 Controller 注册若干个属性编辑器。

注解@ Required 负责检查一个 Bean 在初始化时其 SET 方法是否被执行，如果 SET 方法没有被调用，则 Spring 在解析时会抛出异常来提醒开发者设置对应的属性。@Required 注解只能标注在 SET 方法上，如果将其标注在非 SET 方法上就会被忽略。

Spring 4.0 中引入了条件化配置特性，提供了更加通用的基于条件的 Bean 创建方法；即使用@Conditional 注解。@Conditional 根据满足某一特定条件创建一个特定的 Bean。条件化配置允许配置存在于应用程序中，但在满足某些特定条件之前都忽略这些配置。在 Spring 里可以很方便地编写自定义的条件，只需要实现 Condition 接口并覆盖它的 matches() 方法；如例 3-18 所示。在例 3-18 中，只有当 JdbcTemplateCondition 类的条件成立时才会创建 MyService 这个 Bean。否则，这个 Bean 的声明就会被忽略掉。

【例 3-18】 注解@Conditional 的应用代码示例。

```
//具体条件类，需实现 Condition 接口，并重写 matches(... , ...)方法
public class JdbcTemplateCondition implements Condition {
   @Override
   public boolean matches(ConditionContext context,
                                AnnotatedTypeMetadata metadata) {
        try{
            context.getClassLoader().loadClass(
               "org.springframework.jdbc.core.JdbcTemplate");
            return true;
        } catch (Exception e){
            return false;
        }
   }
}

//声明 bean 时，使用自定义条件类，作为@Conditional 的参数 value
@Conditional(JdbcTemplateCondition.class)
public MyService myService(){...}
}
```

3.2.3 Spring 的注解注入

注解注入是通过注解来实现注入。Spring 中和注入相关的常见注解有@Autowired、

@Resource、@Qualifier、@Service、@Controller、@Repository、@Component 等。其中，注解@Autowired 实现自动注入，注解@Resource 通过指定名称的方式进行注入。注解@Qualifier 和注解@Autowired 配合使用，通过指定名称的方式进行注入。注解@Autowired、@Resource 可以被用来标注字段、构造函数、方法，并进行注入。

注解@Service、@Controller、@Repository 被用来标注类，Spring 扫描注解标注的类时要生成的 Bean。注解@Service、@Controller、@Repository 标注的类分别位于服务层、控制层、数据存储层。注解@Component 是一种泛指，标记被注解的对象是组件。

注解@EnableAspectJAutoProxy 表示开启 AOP 代理自动配置机制，可以通过设置@EnableAspectJAutoProxy(exposeProxy=true)表示使用 AOP 框架来暴露该代理对象，这样 aopContext 就能够访问。从注解@EnableAspectJAutoProxy 的定义可以看出，它引入了一个 AspectJAutoProxyRegister.class 对象，该对象是一个用@EnableAspectJAutoProxy 注解标注的 AnnotationAwareAspectJAutoProxyCreator。

AnnotationAwareAspectJAutoProxyCreator 能通过调用类 AopConfigUtils 的方法 registerAspectJAnnotationAutoProxyCreatorIfNecessary(registry)来注册一个 AOP 代理对象生成器。

注解@Profile 提供了一种隔离应用配置的方法，让这些配置只能在特定环境下生效。任何组件或配置类都能被@Profile 标记，从而限制它们的加载时机。

3.3 Spring Boot 的注解

3.3.1 Spring Boot 基础注解

注解@SpringBootApplication 和注解@Configuration、@EnableAutoConfiguration、@ComponentScan 注解等价。其中，注解@Configuration 标注在类上，等同于 Spring 的 XML 配置文件中 Bean。注解@EnableAutoConfiguration 实现自动配置。注解@ComponentScan 扫描组件，可自动发现和装配 Bean，并把 Bean 加入到程序上下文。

注解@RestController 是注解@Controller 和注解@ResponseBody 的合集，表示被标注的对象是 REST 风格的 Bean，并且是将方法的返回值直接填入 HTTP 响应正文中返回给用户。

注解@JsonBackReference 可以用来解决无限递归调用问题。注解@RepositoryRest-Resource 配合 spring-boot-starter-data-rest 使用，用于创建 RESTful 入口点。注解@Import 用来导入其他配置类。注解@ImportResource 用来加载 XML 配置文件。

注解@Bean 标注方法等价于 XML 配置中的 Bean。注解@Value 注入 Spring Boot 配置文件 application.properties 中配置的属性值。注解@Inject 等价于默认的@Autowired，只是没有 required 属性。

Spring Boot 定义了很多条件，将其运用到了配置类上，这些配置类构成了 Spring Boot 的自动配置。Spring Boot 提供的条件化注解包括@ConditionalOnBean（配置了某个特定 Bean）、@ConditionalOnMissingBean（没有配置特定的 Bean）、@ConditionalOnClass （classpath 目录里有指定的类）、@ConditionalOnMissingClass（classpath 目录里缺少指定的类）、@ConditionalOnExpression（给定的 SpEL 表达式计算结果为 true）、@ConditionalOnJava（Java 的版本匹配特定值或者一个范围值）、@ConditionalOnJndi（参数中给定的 JNDI 位置必须存在一个）、@ConditionalOnProperty（指定的配置属性要有一个明确的值）、@ConditionalOnResource（classpath 目录有指定的资源）、@ConditionalOnWebApplication（是一个 Web 应用程序）、@ConditionalOnNotWebApplication（不是一个 Web 应用程序）。

3.3.2　JPA 注解

注解@Entity 表明被标注的对象是一个实体类，注解@Table(name=" ")指出实体对应的表名；这两个注解一般一起使用。但是如果表名和实体类名相同，则@Table 可以省略。

进行开发项目时，经常会用到将实体类映射到数据库表的操作。有时需要映射的几个实体类有共同的属性，例如编号 ID、创建者、创建时间、备注等。这时，可以把这些属性抽象成一个父类，然后各个实体类继承这个父类。可以使用@MappedSuperclass 注解标注父类，它不会映射到数据库表，但子类在映射时会自动扫描父类的映射属性，并将这些属性添加到子类对应的数据库表中。使用@MappedSuperclass 注解后不能再有@Entity 或@Table 注解。

Spring Data 中提供了很多 DAO 接口，但是依然有可能满足不了日常应用的需要，需要自定义 Repository 实现。注解@NoRepositoryBean 一般用作父类的 Repository，有这个注解 Spring 不会去实例化该 Repository。

注解@Column 标识实体类中属性与数据表中字段的对应关系。如果注解@Column 的字段名与列名相同，则可以省略。@Column 注解一共有 10 个属性，这 10 个属性均为可选属性。常用属性有 name、unique、nullable、table、length、precision、scale 等。其中，name 属性定义了被标注字段在数据库表中所对应字段的名称。unique 属性表示该字段是否为唯一标识，默认值为 false。如果表中有一个字段需要唯一标识，则既可以使用该标记，也可以使用@Table 标记中的@UniqueConstraint。nullable 属性表示该字段是否可以为 null 值，默认值为 true。table 属性定义了包含当前字段的表名。length 属性表示字段的长度，当字段的类型为 varchar 时，该属性才有效；默认值为 255 个字符。precision 属性和 scale 属性表示精度，当字段类型为 double 时，precision 表示数值的总长度，scale 表示小数点所占的位数。

注解@Id 用于声明一个实体类的属性映射为数据库的主键列。该属性通常置于属性声明语句之前，可与声明语句同行，也可写在单独行上。@Id 标注也可置于属性的 getter 方法之前。

注解@GeneratedValue 用于标注主键的生成策略，通过属性 strategy 指定策略。例如，注解@GeneratedValue(strategy = GenerationType.SEQUENCE)表示主键生成策略是 sequence，@GeneratedValue(generator ="repair_seq")指定 sequence 的名字是 repair_seq。在 javax.persistence.GenerationType 中定义了 IDENTITY、AUTO、SEQUENCE 等几种可供选择的策略。其中，IDENTITY 策略表示采用数据库 ID 自增长的方式来自增主键字段，Oracle 不支持这种方式；AUTO SEQUENCE 表示通过序列产生主键，通过@SequenceGenerator

注解指定序列名，MySQL 不支持这种方式。默认情况下，JPA 自动选择一个最适合的底层数据库的主键生成策略，例如 SQL Server 对应的默认策略为 identity，MySQL 对应的默认策略为 auto increment。注解@SequenceGeneretor(name ="repair_seq", sequenceName ="seq_repair", allocationSize = 1)中 name 为 sequence 的名称，以便使用；sequenceName 为数据库的 sequence 名称，两个名称可以一致。

注解@Transient 表示被标注的属性不是一个到数据库表的字段的映射，对象关系映射（Object Relational Mapping，ORM）框架将忽略该属性。如果一个属性并非数据库表的字段映射，就务必将其标示为@Transient；否则，ORM 框架默认其注解为@Basic。注解@Basic(fetch=FetchType.LAZY)可以指定实体属性的加载方式。

注解@JsonIgnore 的作用是 JSON 序列化时将 Java Bean 中的一些属性忽略掉，序列化和反序列化都受影响。例如，如果希望返回的 JSON 数据中不包含属性 goodsInfo 和 extendsInfo 快照值，在实体类快照属性上加注解@JsonIgnore 即可，最后返回的 JSON 数据将不会包含 goodsInfo 和 extendsInfo 两个属性值。

注解@JoinColumn（name="loginId"）表示一张表有指向另一个表的外键。假设 Person 表和 Address 表是一对一的关系，Person 有一个指向 Address 表主键的字段 addressID；可以用注解@JoinColumn 注解 addressID。@OneToOne、@OneToMany、@ManyToOne、@ManyToMany 对应 Hibernate 配置文件中的一对一、一对多、多对一、多对多关系。@ManyToOne 不产生中间表，可以用@JoinColumn（name=" "）来指定外键的名字。@OneToMany 会产生中间表，可以用@OneToMany @JoinColumn（name=" "）避免产生中间表，并且能指定外键的名字。

3.3.3 异常处理注解

注解@ControllerAdvice 包含注解@Component，可以被扫描到，统一处理异常。

注解@ExceptionHandler（Exception.class）用在方法上面表示遇到这个异常就执行所注解的方法。

3.3.4 注解配置解析和使用环境

注解@PreUpdate 用于为相应的生命周期事件指定回调方法。该注解可以应用于实体类、映射超类或回调监听器类的方法。如果要每次更新实体时都要更新属性，可以使用注解@PreUpdate 注释。注解@PreUpdate 不允许更改实体。

注解@PrePersist 帮助在持久化之前自动填充实体属性。@PrePersist 事件在调用 persist()方法后立刻发生，此时的数据还没有真正插入数据库。可以用来在使用 JPA 时记录一些与业务无关的字段，如最后更新时间等。生命周期方法注解（删除没有生命周期事件）@PrePersist 在保存之前被调用，注解 @PostPersist 在保存之后被调用。@PostPersist 事件在数据已经插入数据库后发生。

注解@PostLoad 在 Entity 被映射之后被调用，注解 @EntityListeners 指定外部生命周期事件实现类。注解@PostLoad 事件在执行 EntityManager.find()或 getreference()方法载入

一个实体后、执行 JPA SQL 查询后或 EntityManager.refresh()方法被调用后执行。

注解@PreRemove 和注解@PostRemove 事件的触发由删除实体引起。注解@PreRemove 事件在实体从数据库删除之前触发。注解@PostRemove 事件在实体从数据库中删除后触发。

注解@NoArgsConstructor 提供一个无参的构造方法。注解@AllArgsConstructor 提供一个全参的构造方法。

习题 3

简答题

1．简述 Java 常用注解的含义、用法和功能。
2．简述 Spring 常用注解的含义、用法和功能。
3．简述 Spring Boot 常用注解的含义、用法和功能。

第 4 章

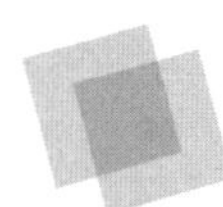

Spring Boot 的 Web 应用开发

Web 应用开发是现代软件开发中重要的一部分。Spring Boot 的 Web 开发内嵌了 Servlet 和服务器，并结合 Spring MVC 来完成开发。本章主要介绍简单的 Web 应用开发，包括如何实现静态 Web 页面，如何实现基于 Thymeleaf 的动态 Web 页面，Thymeleaf 的语法与使用，如何实现基于 Freemarker 的 Web 应用，如何实现对 Ajax 的应用，如何实现 RESTful 风格 Web 应用，如何实现带 Bootstrap 和 jQuery 的 Web 应用，如何实现使用 Servlet、过滤器、监听器和拦截器的 Web 应用开发。

4.1 实现静态 Web 页面

4.1.1 创建类 GreetingController

在第 2 章项目的基础上，在 controller 包中创建一个 GreetingController 类，代码如例 4-1 所示。该代码和 HelloWorldController 文件的作用相同（返回字符串），这是注解@RestController 所起的作用。而注解@GetMapping 和注解@RequestMapping 作用都是用于在 URL 中作为映射信息，区别在于@GetMapping 强调采用 GET 方法获取信息。

【例 4-1】 创建类 GreetingController 的代码示例。

```
package com.bookcode.controller;
import org.springframework.web.bind.annotation.GetMapping;
import org.springframework.web.bind.annotation.RestController;
@RestController                                    //返回的默认结果为字符串
public class GreetingController {
```

```
    @GetMapping("/greeting")                    //映射信息，且采用的是 GET 方法获取信息
    public String greeting() {
          return "greeting";
    }
}
```

注意，一般在创建类、文件之后都需要修改类、文件代码，后面章节用创建类、文件指代创建类、文件和修改代码，示例代码均指修改后的代码。

4.1.2　创建文件 index.html

在项目的 src/main/resources/static 目录下创建文件 index.html，代码如例 4-2 所示。

【例 4-2】 创建文件 index.html 的代码示例。

```
<!DOCTYPE html>
<head>
   <title>Getting Started: Serving Web Content</title>
   <meta http-equiv="Content-Type" content="text/html; charset=UTF-8" />
</head>
<body>
<p>Hello from the static index.html file.</p>
</body>
</html>
```

4.1.3　运行程序

运行程序后，在浏览器中输入 localhost:8888/website/greeting，结果如图 4-1 所示。

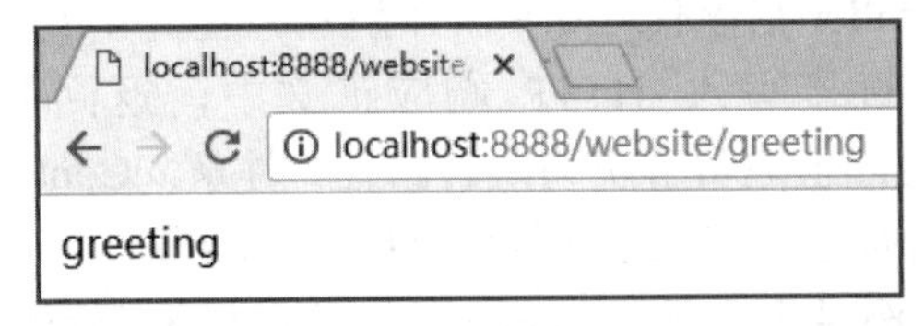

图 4-1　在浏览器中输入 localhost:8888/website/greeting 的结果

在浏览器中输入 localhost:8888/website/index.html 后，结果如图 4-2 所示。

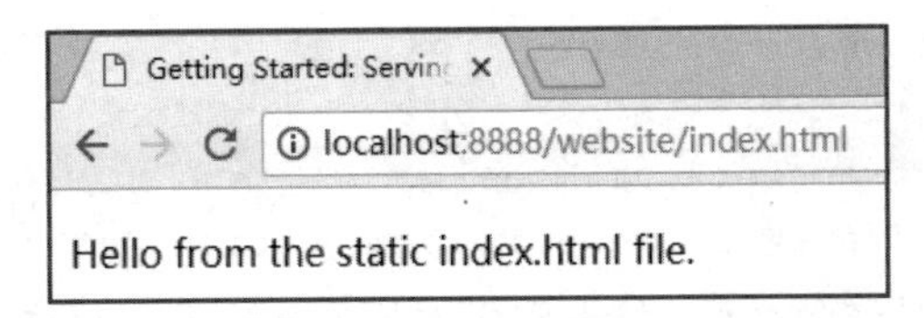

图 4-2　在浏览器中输入 localhost:8888/website/index.html 的结果

index.html 是默认的启动页面，在浏览器中输入 localhost:8888/website/后，结果如图 4-3 所示（和图 4-2 相同）。

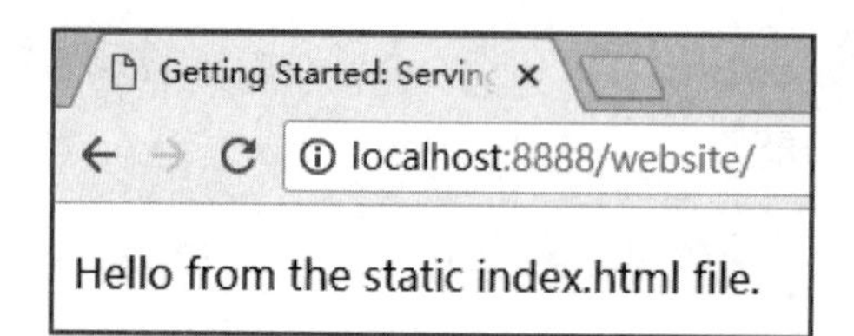

图 4-3　在浏览器中输入 localhost:8888/website/的结果

4.2　实现基于 Thymeleaf 的 Web 应用

视频讲解

在 Spring Boot 中，推荐使用 Thymeleaf 作为动态 Web 页面（视图）的模板引擎。因此，本书示例中的视图主要是基于 Thymeleaf 实现的。

4.2.1　添加依赖

在 pom.xml 文件中，在<dependencies>和</dependencies>之间添加 Thymeleaf 依赖，代码如例 4-3 所示。

【例 4-3】 在 pom.xml 文件中添加 Thymeleaf 依赖的代码示例。

```
<dependency>
    <groupId>org.springframework.boot</groupId>
    <artifactId>spring-boot-starter-thymeleaf</artifactId>
</dependency>
```

注意，后面章节中添加依赖的示例代码均指要添加的代码。

4.2.2　修改类 GreetingController

修改类 GreetingController 后，代码如例 4-4 所示。在例 4-4 中，注解@ResponseBody 表示该方法返回的结果直接写入 HTTP 响应正文中，返回的是字符串。使用注解@RequestMapping 后，返回值通常解析为跳转路径。注解@Controller 返回的是视图，它和注解@Responsebody 组合在一起相当于注解@RestController。注解@RequestParam 表示输入参数 name 信息，在例 4-4 中参数 name 的默认取值为 World，而且参数不是必需的（required=false）。

【例 4-4】 修改类 GreetingController 的代码示例。

```
package com.bookcode.controller;
import org.springframework.stereotype.Controller;
import org.springframework.ui.Model;
import org.springframework.web.bind.annotation.GetMapping;
import org.springframework.web.bind.annotation.RequestParam;
import org.springframework.web.bind.annotation.ResponseBody;
@Controller                     //返回的默认结果为视图，此例子中即是 HTML 文件
public class GreetingController {
    @GetMapping("/greeting")    //映射信息，访问方法为 GET 方法
```

```
    @ResponseBody             //修改了@Controller 的返回要求，返回结果是字符串
    public String greeting() {
        System.out.println("Hello");
        return "greeting";
    }
    @GetMapping("/hi")        //映射信息，访问方法为 GET 方法
public String hi(@RequestParam(name="name", required=false, defaultValue=
"World")String name, Model model) {
        model.addAttribute("name", name);
        return "hi";          //返回的是视图,即返回位于 templates 目录下的 hi.html
    }
}
```

4.2.3　创建文件 hi.html

在项目目录 src/main/resources/templates 下创建文件 hi.html，代码如例 4-5 所示。

【例 4-5】 创建文件 hi.html 的代码示例。

```
<!DOCTYPE html>
<!--thymeleaf 的命名空间-->
<html xmlns:th="http://www.thymeleaf.org">
<head>
    <title>Getting Started: Serving Web Content</title>
    <meta http-equiv="Content-Type" content="text/html; charset=UTF-8" />
</head>
<body>
<!--${name}获取 GreetingController 类中的 name 变量信息并输出-->
<p th:text="'Hello, ' + ${name} + '!'" />
</body>
</html>
```

4.2.4　运行程序

运行程序后在浏览器中输入 localhost:8080/hi?name=zs，结果如图 4-4 所示。请注意将端口改为 8080。

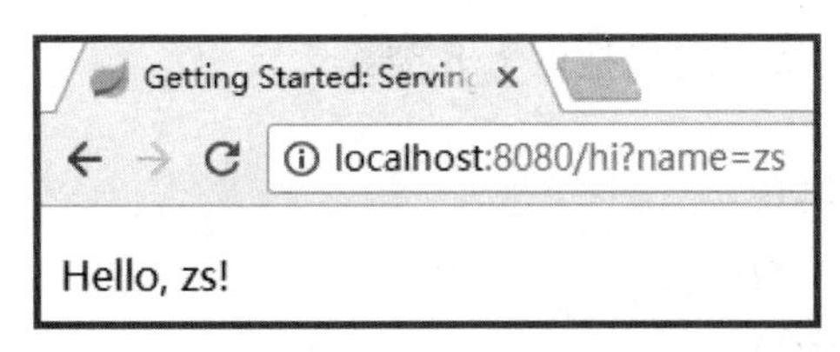

图 4-4　在浏览器中输入 localhost:8080/hi?name=zs 的结果

在浏览器中输入 localhost:8080/hi，显示结果如图 4-5 所示。此时，参数 name 将选择默认值 World。关于参数 name 的默认值请结合例 4-4 的代码来分析。

图 4-5 在浏览器中输入 localhost:8080/hi 的运行结果

4.3 Thymeleaf 的语法与使用

由于 Spring Boot 开发中推荐使用的视图是基于 Thymeleaf 的，本节简要介绍 Thymeleaf 的语法与使用。

4.3.1 Thymeleaf 基础知识

Thymeleaf 是一个模板引擎，以便显示由应用程序生成的数据或文本。它适合在 Web 应用程序中为 HTML 5 提供服务，也可以处理任何 XML 文件。Thymeleaf 具有“开箱即用”的特点；允许处理 XML、HTML、JavaScript、CSS、普通文本等模板，每种模板都称为模板模式。

Thymeleaf 命名空间被声明为 th:*属性，代码如例 4-6 所示。

【例 4-6】 Thymeleaf 命名空间的声明代码示例。

```
<html  xmlns:th="http://www.thymeleaf.org">
```

对文本进行外部化是将模板代码片段从模板文件中提取出来，以便将它们保存在特定的单独文件（通常是.properties 文件）中，并且可以用其他语言（称为国际化或简称 i18n）编写的等效文本替换它们。外部化的文本片段通常称为消息。消息总是有一个标识它们的键。Thymeleaf 允许指定一个文本以对应于一个特定的消息，#{…}语法如例 4-7 所示。

【例 4-7】 Thymeleaf 中#{…}语法的应用代码示例。

```
<p th:text="#{home.welcome}">Welcome to our grocery store!</p>
```

4.3.2 Thymeleaf 的标准表达式

Thymeleaf 的标准表达式主要包括以下 8 类。

1．简单表达式

（1）变量表达式：${…}。

（2）选择变量表达式：*{…}。

（3）消息表达式：#{…}。

（4）链接 URL 表达式：@{…}。这里 URL 包括绝对网址（如 http://www.thymeleaf.org）和相对网址（如 user/login.html）。

（5）分段表达式：~{…}。

2．字面量

（1）文本：'one text'，'Another one!'等。

（2）数值：0，4，3.0，12.4 等。

（3）布尔值：true，false。

（4）空： null。

（5）标记：one，sometext，other 等。

3．文本操作

（1）字符串连接： +。

（2）文本替换： |The name is ${name}|。

4．算术运算

（1）二元运算：+，–，*，/，%等。

（2）一元运算：–。

5．布尔运算

（1）二元运算：and，or。

（2）布尔否定（一元运算）： !，not。

6．比较和等价

（1）比较：>，<，>=，<=，gt，lt，ge，le 等。

（2）等价：==，!=，eq，ne 等。

7．条件运算操作

（1）IF-THEN： (if) ? (then)。

（2）IF-THEN-ELSE： (if) ? (then) : (else)。

（3）取默认值： (value) ?: (defaultvalue)。

8．特殊标记

无操作：_。

所有这些功能可以组合和嵌套，例 4-8 涵盖了上述大部分表达式。

【例 4-8】 Thymeleaf 标准表达式的应用代码示例。

```
'User is of type ' + (${user.isAdmin()} ? 'Administrator' : (${user.type} ?:
'Unknown'))
```

4.3.3　Thymeleaf 的表达式对象

可以使用${…}语法表示一个变量表达式的值。例 4-9 的代码包含一个名为 OGNL（对象图导航语言）的语言表达式，将在上下文变量映射上执行。

【例 4-9】 Thymeleaf 中${…}语法的应用代码示例。

```
<p>Today is: <span th:text="${today}">13 February 2011</span></p>
```

变量表达式不仅可以写成${…}表达式，还可以写入表达式*{…}中。不过，星号（*）表示所选对象上的表达式，而不是整个上下文变量的映射。

在上下文变量上使用 OGNL 表达式时，一些对象可用于表达式以获得更大的灵活性。

这些对象将被引用，从#符号开始。

#ctx 表示上下文对象。

#vars 表示上下文变量。

#locale 表示上下文语言环境。

#request 表示（只在 Web 上下文中）HttpServletRequest 对象。

#response 表示（只在 Web 上下文中）HttpServletResponse 对象。

#session 表示（只在 Web 上下文中）HttpSession 对象。

#servletContext 表示（只在 Web 上下文中）ServletContext 对象。

示例代码如例 4-10 所示。

【例 4-10】 #的应用代码示例。

```
国家：<span th:text="${#locale.country}">US</span>
```

除了这些基本的对象之外，Thymeleaf 还提供了一套实用对象，帮助实现在表达式中执行常见任务。

#dates 表示 java.util.Date 对象的实用方法（格式化，组件提取等）。

#calendars 表示 java.util.Calendars 对象的实用方法。

#numbers 表示格式化数字对象的实用方法。

#strings 表示 String 对象的实用方法（contains、startsWith、prepending、appending 等）。

#objects 表示一般对象的实用方法。

#bools 表示布尔评估的实用方法。

#arrays 表示数组的实用方法。

#lists 表示列表的实用方法。

#sets 表示集合的实用方法。

#maps 表示地图的实用方法。

#aggregates 表示用于在数组或集合上创建聚合的实用方法。

#messages 表示用于在变量表达式中获得外部消息的实用方法，与使用#{…}语法获得的方式相同。

#ids 表示用于处理可能重复的 id 属性的实用方法（例如，作为迭代的结果）。

代码如例 4-11 所示。

【例 4-11】 Thymeleaf 中实用对象的应用代码示例。

```
今天是：<span th:text="${#calendars.format(today,'dd MMMM yyyy')}">5 May
2018</span>
```

4.3.4 Thymeleaf 设置属性

可以使用 th:*任务设置特定标签属性的属性值（而不仅仅是任何属性 th:attr），这些属性如下所示。

th:abbr	th:accept	th:accept-charset
th:accesskey	th:action	th:align
th:alt	th:archive	th:audio
th:autocomplete	th:axis	th:background
th:bgcolor	th:border	th:cellpadding
th:cellspacing	th:challenge	th:charset
th:cite	th:class	th:classid
th:codebase	th:codetype	th:cols
th:colspan	th:compact	th:content
th:contenteditable	th:contextmenu	th:data
th:datetime	th:dir	th:draggable
th:dropzone	th:enctype	th:for
th:form	th:formaction	th:formenctype
th:formmethod	th:formtarget	th:frame
th:frameborder	th:headers	th:height
th:high	th:href	th:hreflang
th:hspace	th:http-equiv	th:icon
th:id	th:keytype	th:kind
th:label	th:lang	th:list
th:longdesc	th:low	th:manifest
th:marginheight	th:marginwidth	th:max
th:maxlength	th:media	th:method
th:min	th:name	th:optimum
th:pattern	th:placeholder	th:poster
th:preload	th:radiogroup	th:rel
th:rev	th:rows	th:rowspan
th:rules	th:sandbox	th:scheme
th:scope	th:scrolling	th:size
th:sizes	th:span	th:spellcheck
th:src	th:srclang	th:standby
th:start	th:step	th:style
th:summary	th:tabindex	th:target
th:title	th:type	th:usemap
th:value	th:valuetype	th:vspace
th:width	th:wrap	th:xmlbase
th:xmllang	th:xmlspace	

有两个比较特殊的属性为 th:alt-title（将同时设置 alt 和 title 属性）和 th:lang-xmllang（将同时设置 lang 和 xml:lang 属性）。

另外，存在一些具有固定值的布尔属性，如下所示。没有值的属性意味着属性值为 true。

th:async th:autofocus th:autoplay
th:checked th:controls th:declare
th:default th:defer th:disabled
th:formnovalidate th:hidden th:ismap
th:loop th:multiple th:novalidate
th:nowrap th:open th:pubdate
th:readonly th:required th:reversed
th:scoped th:seamless th:selected

4.3.5 Thymeleaf 的迭代和条件语句

为了在 Web 页面上列出某个企业生产的全部产品，需要用到表格，每个产品都将显示在一行。所以，Thymeleaf 需要有迭代功能，Thymeleaf 为此提供了一个属性 th:each。Thymeleaf 用 th:each 提供了一个有用的机制来跟踪具有迭代状态的状态变量。状态变量在一个 th:each 属性中定义并包含以下变量。

（1）index 表示当前迭代索引，从 0 开始计数。

（2）count 表示当前迭代索引，从 1 开始计数。

（3）size 表示迭代变量中的元素总数。

（4）current 表示每个迭代的 iter 变量。

（5）布尔值 even/odd 表示目前的迭代是偶数还是奇数。

（6）布尔值 first 表示目前的迭代是否是第一个。

（7）布尔值 last 表示目前的迭代是否是最后一个。

代码如例 4-12 所示。

【例 4-12】 Thymeleaf 中迭代语句应用的代码示例。

```
<table>
  <tr>
    <th>产品名称</th>
    <th>产品价格</th>
    <th>产品库存</th>
  </tr>
  <tr th:each="prod, iterStat : ${prods}" th:class="${iterStat.odd}? 'odd'">
    <td th:text="${prod.name}">Onions</td>
    <td th:text="${prod.price}">2.41</td>
    <td th:text="${prod.inStock}? #{true} : #{false}">yes</td>
  </tr>
</table>
```

条件语句和一般编程语言类似，代码如例 4-13 所示。

【例 4-13】 Thymeleaf 中条件语句应用的代码示例。

```
<a href="comments.html"
```

```
    th:href="@{/product/comments(prodId=${prod.id})}"
    th:if="${not #lists.isEmpty(prod.comments)}">view</a>
<a href="comments.html"
    th:href="@{/comments(prodId=${prod.id})}"
    th:unless="${#lists.isEmpty(prod.comments)}">view</a>
<div th:switch="${user.role}">
  <p th:case="'admin'">User is an administrator</p>
  <p th:case="#{roles.manager}">User is a manager</p>
</div>
```

4.3.6 Thymeleaf 模板片段的定义和引用

因为 Jackson JSON 处理库位于类路径 classpath 中，模板 RestTemplate 将使用它（通过消息转换器）将传入的 JSON 数据转换为 Quote 对象。接着，Quote 对象的内容被作为日志信息发送到控制台。

在 Web 中可能需要在模板中包含其他模板片段，常见的用途是在不同的页面中添加固定的页脚、标题、菜单等片段。为了做到这一点，Thymeleaf 需要先定义可用于包含的片段，可以通过使用 th:fragment 属性来完成。例如，将一个版权页脚添加到所有某个 Web 页面时，需要先定义一个版权页脚片段（footer.html），代码如例 4-14 所示。

【例 4-14】 Thymeleaf 中模板片段定义的代码示例。

```
<!DOCTYPE html SYSTEM "http://www.thymeleaf.org/dtd/xhtml1-strict-
thymeleaf-4.dtd">
<html xmlns="http://www.w3.org/1999/xhtml" xmlns:th="http://www.thymeleaf.org">
  <body>
    <div th:fragment="copy">
      &copy; 2011 The Good Thymes Virtual Grocery
    </div>
  </body>
</html>
```

接着，可以在 Web 页面中使用 th:insert、th:include 或者 th:replace 属性来包含页脚片段，代码如例 4-15 所示。

【例 4-15】 Thymeleaf 中引用模板片段的代码示例。

```
<body>
...
<div th:insert="footer :: copy"></div>
<div th:include="footer :: copy"></div>
<div th:replace="footer :: copy"></div>
</body>
```

th:insert 将用指定片段替换主标签，th:include 将用片段的实际内容替换主标签，th:replace 将用实际片段替换主标签。

4.4 实现基于 Freemarker 的 Web 应用

视频讲解

Freemarker 是一款模板引擎，用来生成输出文本（HTML 网页、电子邮件、配置文件、源代码等）的通用工具。它不是面向最终用户的，而是一个 Java 类库，是一款程序员可以嵌入他们所开发产品的组件。Freemarker 模板编写语言（Freemarker Template Language，FTL）是简单、专用的语言。在模板中主要关注如何展现数据，而在模板外关注要展示什么数据。Thymeleaf 和 Freemarker 在一个项目中能起到相同的作用，即作为视图显示结果，而且它们可以在一个项目中共存。

4.4.1 添加依赖

在 pom.xml 文件中，在<dependencies>和</dependencies>之间添加 Freemarker 依赖，代码如例 4-16 所示。

【例 4-16】 添加 Freemarker 依赖的代码示例。

```
<dependency>
    <groupId>org.springframework.boot</groupId>
    <artifactId>spring-boot-starter-freemarker</artifactId>
 </dependency>
```

4.4.2 创建类 TemplateController

在包 com.bookcode 下创建子包 controller，并在 com.bookcode.controller 包中创建类 TemplateController，代码如例 4-17 所示。

【例 4-17】 创建类 TemplateController 的代码示例。

```
package com.bookcode.controller;
import org.springframework.stereotype.Controller;
import org.springframework.web.bind.annotation.RequestMapping;
import java.util.Map;
@Controller
public class TemplateController {
    @RequestMapping("/helloFtl")
    public String helloFtl(Map<String,Object> map){
        map.put("hello","基于 Freemarker from TemplateController.helloFtl");
        return"/helloFtl";
    }
}
```

4.4.3 创建文件 helloFtl.ftl

在 resources/templates 目录下，创建文件 helloFtl.ftl，代码如例 4-18 所示。

【例 4-18】 创建文件 helloFtl.ftl 的代码示例。

```
<!DOCTYPE html>
<html xmlns="http://www.w3.org/1999/xhtml">
<head>
   <title>Hello World!</title>
</head>
<body>
<h1>Hello.v.2</h1>
<p>${hello}</p>
</body>
</html>
```

4.4.4　运行程序

运行程序后，在浏览器中输入 127.0.0.1:8080/helloFtl，结果如图 4-6 所示。

图 4-6　在浏览器中输入 127.0.0.1:8080/helloFtl 的结果

4.5　Spring Boot 对 Ajax 的应用

视频讲解

Ajax 即 Asynchronous JavaScript And XML（异步 JavaScript 和 XML），是指一种创建快速、动态、交互式 Web 应用的开发技术。不使用 Ajax 的传统网页如果需要更新内容，则必须重载整个页面。通过在后台与服务器进行少量数据交换，Ajax 可以使网页实现异步更新。这意味着可以在不重新加载整个网页的情况下，对网页的某部分进行更新。

4.5.1　创建类 HelloWorldAjaxController

在包 com.bookcode.controller 中创建类 HelloWorldAjaxController，代码如例 4-19 所示。

【例 4-19】 创建类 HelloWorldAjaxController 的代码示例。

```
package com.bookcode.controller;
import org.springframework.web.bind.annotation.RequestMapping;
import org.springframework.web.bind.annotation.RestController;
@RestController
```

```
@RequestMapping("/ajax")
public class HelloWorldAjaxController {
    @RequestMapping("/hello")
    public String say(){
        return "{'message1':'SpringBoot你好','message2','你好Ajax'}";
    }
}
```

4.5.2 创建文件 index.html

在 resources/static 目录下创建文件 index.html，代码如例 4-20 所示。

【例 4-20】 创建文件 index.html 的代码示例。

```
<!DOCTYPE html>
<html>
<head>
    <meta charset="UTF-8">
    <title>Insert title here</title>
    <script src="http://www.java1234.com/jquery-easyui-1.3.3/jquery.min.js"></script>
    <script type="text/javascript">
        function show(){
            $.post("ajax/hello",{},function(result){
                alert(result);
            });
        }
    </script>
</head>
<body>
<button onclick="show()">Ajax测试按钮</button>
</body>
</html>
```

4.5.3 运行程序

运行程序后，在浏览器中输入 localhost:8080/index.html；单击“Ajax 测试按钮”按钮，弹出对话框，结果如图 4-7 所示。

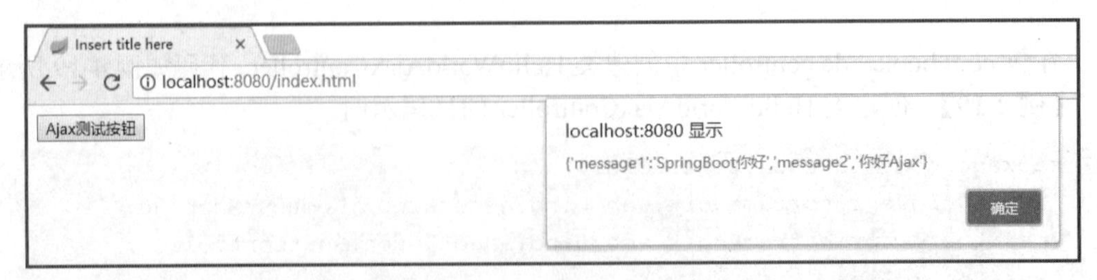

图 4-7 在浏览器中输入 localhost:8080/index.html 后单击“Ajax 测试按钮”按钮的结果

4.6　Spring Boot 实现 RESTful 风格 Web 应用

视频讲解

REST（Representational State Transfer）描述了一个架构样式的网络系统，如 Web 应用程序。REST 指的是一组架构约束条件和原则，满足这些约束条件和原则的应用程序或设计就是 RESTful。Web 应用程序最重要的 REST 原则是客户端和服务器之间的交互在请求之间是无状态的。从客户端到服务器的每个请求都必须包含理解请求所必需的信息。此外，无状态请求可以由任何可用服务器回答，客户端可以缓存数据以改进性能。服务器端的应用程序状态和功能可以分为各种资源，常见的资源包括应用程序对象、数据库记录、算法等。每个资源都使用 URI（Universal Resource Identifier）得到一个唯一的地址。所有资源都共享统一的接口，以便在客户端和服务器之间传输状态。传输使用的方法是标准的 HTTP 方法，如 GET、PUT、POST 和 DELETE。

4.6.1　创建类 BlogController

在包 com.bookcode.controller 中创建类 BlogController，代码如例 4-21 所示。

【例 4-21】 创建类 BlogController 的代码示例。

```
package com.bookcode.controller;
import org.springframework.stereotype.Controller;
import org.springframework.web.bind.annotation.PathVariable;
import org.springframework.web.bind.annotation.RequestMapping;
import org.springframework.web.bind.annotation.RequestParam;
import org.springframework.web.servlet.ModelAndView;
@Controller
@RequestMapping("/blog")
public class BlogController {
    @RequestMapping("/{id}")
    public ModelAndView show(@PathVariable("id") Integer id){
        ModelAndView mav=new ModelAndView();
        mav.addObject("id", id);
        mav.setViewName("blog");
        return mav;
    }
    @RequestMapping("/query")
    public ModelAndView query(@RequestParam(value="q",required=false)String q){
        ModelAndView mav=new ModelAndView();
        mav.addObject("q", q);
        mav.setViewName("query");
        return mav;
```

```
    }
}
```

4.6.2 创建文件 index.html

在 resources/static 目录下创建文件 index.html，代码如例 4-22 所示。

【例 4-22】 创建文件 index.html 的代码示例。

```
<!DOCTYPE html>
<html>
<head>
    <meta charset="UTF-8">
    <title>Insert title here</title>
    <script src="http://www.java1234.com/jquery-easyui-1.3.3/jquery.min.js">
    </script>
    <script type="text/javascript">
        function show(){
            $.post("ajax/hello",{},function(result){
                alert(result);
            });
        }
    </script>
</head>
<body>
<button onclick="show()">Ajax 测试按钮</button><br/>
<a href="/blog/21">博客</a><br/>
<a href="/blog/query?q=123456">搜索</a>
</body>
</html>
```

4.6.3 创建文件 blog.html

在 resources/templates 目录下创建文件 blog.html，代码如例 4-23 所示。

【例 4-23】 创建文件 blog.html 的代码示例。

```
<!DOCTYPE html>
<html xmlns:th="http://www.thymeleaf.org">
<head>
    <title>Getting Started: Serving Web Content</title>
    <meta http-equiv="Content-Type" content="text/html; charset=UTF-8" />
</head>
<body>
<p th:text="'博客编号:' + ${id} + '!'" />
```

```
</body>
</html>
```

4.6.4　创建文件 query.html

在 resources/templates 目录下创建文件 query.html，代码如例 4-24 所示。

【例 4-24】 创建文件 query.html 的代码示例。

```
<!DOCTYPE html>
<html xmlns:th="http://www.thymeleaf.org">
<head>
    <title>Getting Started: Serving Web Content</title>
    <meta http-equiv="Content-Type" content="text/html; charset=UTF-8" />
</head>
<body>
<p th:text="'查询, ' + ${q} + '!'" />
</body>
</html>
```

4.6.5　运行程序

运行程序后，在浏览器中输入 localhost:8080/index.html，结果如图 4-8 所示。单击“博客”链接后，结果如图 4-9 所示。单击“搜索”链接后，结果如图 4-10 所示。对比例 4-21 中类 BlogController 的代码和例 4-19，可以发现 BlogController 类没有实现访问路径 /ajax/hello 对应的方法，因此单击图 4-8 中的“Ajax 测试按钮”按钮后没有任何反应。

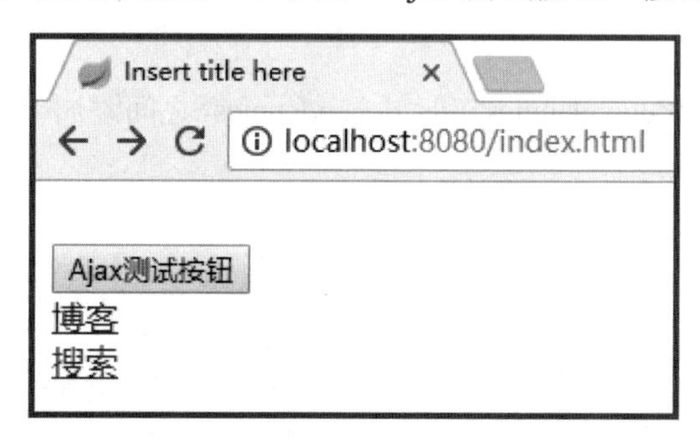

图 4-8　在浏览器中输入 localhost:8080/index.html 的结果

图 4-9　单击“博客”链接后的结果

图 4-10 单击“搜索”链接后的结果

4.7 带 Bootstrap 和 jQuery 的 Web 应用

视频讲解

Bootstrap 是比较受欢迎的前端组件库之一，可以用于开发响应式布局、移动设备优先的 Web 项目。Bootstrap 是一套用于 HTML、CSS 和 JavaScript 开发的开源工具集。Bootstrap 中包含了丰富的 Web 组件，利用这些组件可以快速地搭建一个漂亮、功能完备的网站。Bootstrap 中包含的组件有下拉菜单、按钮组、按钮下拉菜单、导航、导航条、路径导航、分页、排版、缩略图、警告对话框、进度条、媒体对象等。

jQuery 是一个快速、简洁的 JavaScript 框架，是继 Prototype 之后又一个优秀的 JavaScript 代码库（或 JavaScript 框架）。jQuery 设计的宗旨是“Write Less，Do More”，即倡导写更少的代码做更多的事情。它封装了 JavaScript 常用的功能代码，提供一种简便的 JavaScript 设计模式，优化了 HTML 文档操作、事件处理、动画设计和 Ajax 交互。jQuery 的核心特性可以总结为：具有独特的链式语法和短小清晰的多功能接口；具有高效灵活的 CSS 选择器，并且可对 CSS 选择器进行扩展；拥有便捷的插件扩展机制和丰富的插件；兼容各种主流浏览器。

4.7.1 添加依赖

在 pom.xml 文件中<dependencies>和</dependencies>之间添加 Web 和 Thymeleaf 依赖，代码如例 4-25 所示。

【例 4-25】 添加 Web 和 Thymeleaf 依赖的代码示例。

```
<dependency>
    <groupId>org.springframework.boot</groupId>
    <artifactId>spring-boot-starter-web</artifactId>
</dependency>
<dependency>
    <groupId>org.springframework.boot</groupId>
    <artifactId>spring-boot-starter-thymeleaf</artifactId>
 </dependency>
```

4.7.2 创建类 Person

在包 com.bookcode.entity 中创建类 Person，代码如例 4-26 所示。

【例 4-26】　创建类 Person 的代码示例。

```
package com.bookcode.entity;
public class Person {
    private String name;
    private Integer age;
    private String address;
    public Person() {
        super();    }
    public Person(String name, Integer age, String address) {
        super();
        this.name = name;
        this.age = age;
        this.address = address;
    }
    public String getName() {
        return name;
    }
    public void setName(String name) {
        this.name = name;
    }
    public Integer getAge() {
        return age;
    }
    public void setAge(Integer age) {
        this.age = age;
    }
    public String getAddress() {
        return address;
    }
    public void setAddress(String address) {
        this.address = address;
    }
}
```

4.7.3　创建类 BJController

在包 com.bookcode.controller 中创建类 BJController，代码如例 4-27 所示。

【例 4-27】　创建类 BJController 的代码示例。

```
package com.bookcode.controller;
import com.bookcode.entity.Person;
import org.springframework.stereotype.Controller;
import org.springframework.ui.Model;
import org.springframework.web.bind.annotation.RequestMapping;
```

```
import java.util.ArrayList;
import java.util.List;
@Controller
public class BJController {
    @RequestMapping("/")
    public String index(Model model){
        Person single = new Person("张三",11,"徐州");
        List<Person> people = new ArrayList<Person>();
        Person p1 = new Person("李四",11,"江苏");
        Person p2 = new Person("王五",22,"湖北");
        Person p3 = new Person("钱进",33,"北京");
        people.add(p1);
        people.add(p2);
        people.add(p3);
        model.addAttribute("singlePerson", single);
        model.addAttribute("people", people);
        return "index";
    }
}
```

4.7.4 添加辅助文件

下载 Bootstrap、jQuery 并将它们添加到项目中 resources/static 目录下，如图 4-11 所示。

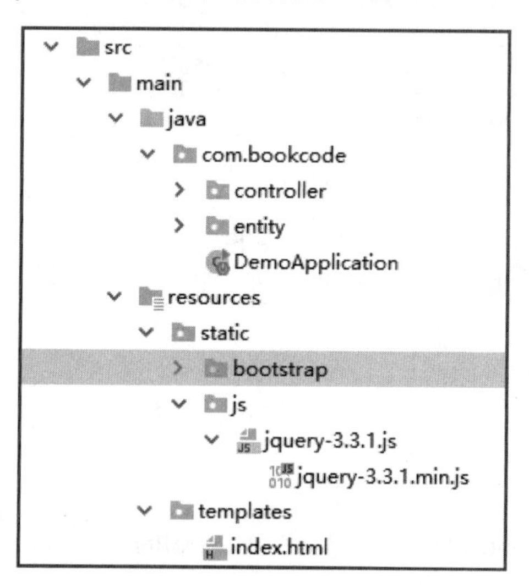

图 4-11 在项目中添加 Bootstrap、jQuery 等辅助文件后的结果

4.7.5 创建文件 index.html

在 resources/templates 目录下创建文件 index.html，代码如例 4-28 所示。

【例 4-28】 创建文件 index.html 的代码示例。

```
<!DOCTYPE html>
<html xmlns:th="http://www.thymeleaf.org">
<head>
    <meta content="text/html;charset=UTF-8"/>
    <meta http-equiv="X-UA-Compatible" content="IE=edge"/>
    <meta name="viewport" content="width=device-width, initial-scale=1"/>
    <link th:href="@{bootstrap/css/bootstrap.css}" rel="stylesheet"/>
</head>
<body>
<div class="panel panel-primary">
    <div class="panel-heading">
        <h3 class="panel-title">访问 model</h3>
    </div>
    <div class="panel-body">
        <span th:text="${singlePerson.name}"></span>
    </div>
</div>
<div th:if="${not #lists.isEmpty(people)}">
    <div class="panel panel-primary">
        <div class="panel-heading">
            <h3 class="panel-title">列表</h3>
        </div>
        <div class="panel-body">
            <ul class="list-group">
                <li class="list-group-item" th:each="person:${people}">
                    <span th:text="${person.name}"></span>
                    <span th:text="${person.age}"></span>
                    <span th:text="${person.address}"></span>
          <button class="btn" th:onclick="'getName(\'' + ${person.name} +
          '\');'">获得名字</button>
                </li>
            </ul>
        </div>
    </div>
</div>
<script th:src="@{js/jquery-3.3.1.min.js}" type="text/javascript"></script>
<!-- 2 -->
<script th:src="@{bootstrap/js/bootstrap.min.js}"></script><!-- 2 -->
<script th:inline="javascript">
    var single = [[${singlePerson}]];
    console.log(single.name+"/"+single.age+"/"+single.address)
```

```
    function getName(name){
        console.log(name);
    }
</script>
</body>
</html>
```

4.7.6 运行程序

运行程序后，在浏览器中输入 localhost:8080，并单击“获得名字”按钮后，结果如图 4-12 所示。

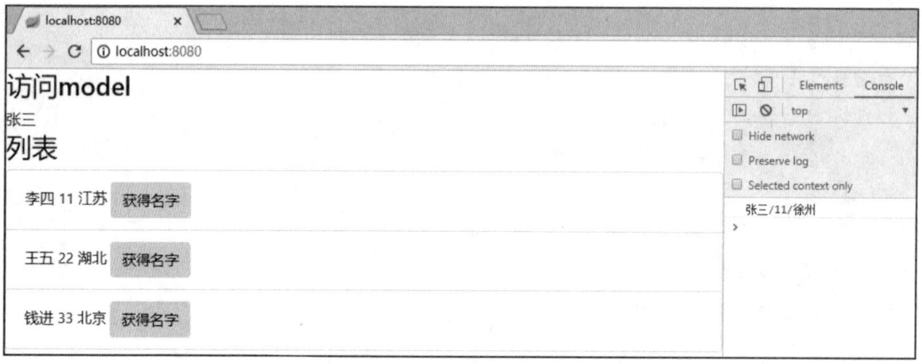

图 4-12 在浏览器中输入 localhost:8080 并单击“获得名字”按钮后的结果

4.8 使用 Servlet、过滤器、监听器和拦截器

视频讲解

Java Servlet（Server Applet）是用 Java 编写的服务器端程序，主要功能在于交互式地浏览和修改数据并生成动态 Web 内容。狭义的 Servlet 是指 Java 语言实现的一个接口，广义的 Servlet 是指任何实现了这个 Servlet 接口的类，一般情况下人们将 Servlet 理解为后者。Servlet 运行在支持 Java 的应用服务器中。从理论上讲，Servlet 可以响应任何类型的请求，但绝大多数情况下 Servlet 只用来扩展基于 HTTP 的 Web 服务器。

过滤器依赖于 Servlet 容器，是 Java EE 标准；在请求进入容器之后还未进入 Servlet 之前进行预处理，并且在请求结束返回给前端之前进行后期处理。过滤器的实现基于函数回调，它可以对几乎所有请求进行过滤。其缺点是一个过滤器实例只能在容器初始化时调用一次。使用过滤器的目的是做一些过滤操作或修改 HttpServletRequest 的参数，如修改字符编码、过滤低俗文字和危险字符等。

监听器是一个实现特定接口的普通 Java 程序，这个程序专门用于监听另一个 Java 对象的方法调用或者属性改变。当被监听对象发生上述事件后，监听器的某个方法将立即执行。

拦截器不依赖于 Servlet 容器，依赖于具体的 Web 框架，在 Spring MVC 中就是依赖于 Spring MVC 框架。在实现上基于 Java 的反射机制，属于面向切面编程（AOP）的一种运用。由于拦截器是基于 Web 框架的调用，因此可以使用 Spring 的依赖注入（DI）获取 IoC 容器中的各个 Bean，进行业务操作。一个拦截器实例在一个控制器生命周期之内可以多次调用，但是只能对控制器请求进行拦截。

4.8.1　创建类 MyServlet1

创建类 MyServlet1，代码如例 4-29 所示。

【例 4-29】 创建类 MyServlet1 的代码示例。

```
package com.bookcode.servlet;
import javax.servlet.ServletException;
import javax.servlet.http.HttpServlet;
import javax.servlet.http.HttpServletRequest;
import javax.servlet.http.HttpServletResponse;
import java.io.IOException;
import java.io.PrintWriter;
public class MyServlet1 extends HttpServlet {
    @Override
    protected void doGet(HttpServletRequest req, HttpServletResponse resp)
    throws ServletException,
            IOException {
        System.out.println(">>>>>>>>>>doGet()<<<<<<<<<<<");
        doPost(req, resp);
    }
    @Override
    protected void doPost(HttpServletRequest req, HttpServletResponse resp)
    throws ServletException,
            IOException {
        System.out.println(">>>>>>>>>>doPost()<<<<<<<<<<<");
        resp.setContentType("text/html");
        resp.setCharacterEncoding("utf-8");
        PrintWriter out = resp.getWriter();
        out.println("<html>");
        out.println("<head>");
        out.println("<title>Hello World</title>");
        out.println("</head>");
        out.println("<body>");
        out.println("<h1>这是: MyServlet1</h1>");
        out.println("</body>");
```

```
        out.println("</html>");
    }
}
```

4.8.2 修改入口类 1

修改入口类，代码如例 4-30 所示。

【例 4-30】 修改入口类 DemoApplication 的代码示例。

```
package com.bookcode;
import com.bookcode.servlet.MyServlet1;
import org.springframework.boot.SpringApplication;
import org.springframework.boot.autoconfigure.SpringBootApplication;
import org.springframework.boot.web.servlet.ServletRegistrationBean;
import org.springframework.context.annotation.Bean;
@SpringBootApplication
public class DemoApplication {
    @Bean
    public ServletRegistrationBean MyServlet1(){
        return new ServletRegistrationBean(new MyServlet1(),"/myServlet/*");
    }
    public static void main(String[] args) {
        SpringApplication.run(DemoApplication.class, args);
    }
}
```

4.8.3 运行程序 1

运行程序后，在浏览器中输入 localhost:8080/myServlet，结果如图 4-13 所示。

图 4-13 在浏览器中输入 localhost:8080/myServlet 后的结果

4.8.4 创建类 MyServlet2

创建类 MyServlet2，代码如例 4-31 所示。

【例 4-31】 创建类 MyServlet2 的代码示例。

```
package com.bookcode.servlet;
import javax.servlet.ServletException;
import javax.servlet.annotation.WebServlet;
import javax.servlet.http.HttpServlet;
import javax.servlet.http.HttpServletRequest;
import javax.servlet.http.HttpServletResponse;
import java.io.IOException;
import java.io.PrintWriter;
@WebServlet(urlPatterns="/myServlet2/*", description="Servlet 的说明")
public class MyServlet2 extends HttpServlet {
    @Override
    protected void doGet(HttpServletRequest req, HttpServletResponse resp)
    throws ServletException,
            IOException {
        System.out.println(">>>>>>>>>>doGet()<<<<<<<<<<<");
        doPost(req, resp);
    }
    @Override
    protected void doPost(HttpServletRequest req, HttpServletResponse resp)
    throws ServletException,
            IOException {
        System.out.println(">>>>>>>>>>doPost()<<<<<<<<<<<");
        resp.setContentType("text/html");
        resp.setCharacterEncoding("utf-8");
        PrintWriter out = resp.getWriter();
        out.println("<html>");
        out.println("<head>");
        out.println("<title>Hello World</title>");
        out.println("</head>");
        out.println("<body>");
        out.println("<h1>这是: MyServlet2</h1>");
        out.println("</body>");
        out.println("</html>");
}
}
```

4.8.5　修改入口类 2

修改入口类，代码如例 4-32 所示。

【例 4-32】 修改入口类的代码示例。

```
package com.bookcode;
import org.springframework.boot.SpringApplication;
```

```
import org.springframework.boot.autoconfigure.SpringBootApplication;
import org.springframework.boot.web.servlet.ServletComponentScan;
@SpringBootApplication
@ServletComponentScan
public class DemoApplication {
    public static void main(String[] args) {
        SpringApplication.run(DemoApplication.class, args);
    }
}
```

4.8.6 运行程序 2

运行程序后，在浏览器中输入 localhost:8080/myServlet2，结果如图 4-14 所示。对比例 4-30 和例 4-32 代码，可以发现：前者是用@Bean 注解注入 Servlet 类，而后者采用@ServletComponentScan 注解自动扫描 Servlet 类。

图 4-14 在浏览器中输入 localhost:8080/myServlet2 的结果

4.8.7 创建类 MyFilter

创建类 MyFilter，代码如例 4-33 所示。

【例 4-33】 创建类 MyFilter 的代码示例。

```
package com.bookcode.servlet;
import javax.servlet.*;
import javax.servlet.annotation.WebFilter;
import java.io.IOException;
@WebFilter(filterName="myFilter",urlPatterns="/*")
public class MyFilter implements Filter {
@Override
public void init(FilterConfig config) throws ServletException {
      System.out.println("过滤器初始化");        }
@Override
public void doFilter(ServletRequest request, ServletResponse response,
                 FilterChain chain) throws IOException, ServletException {
      System.out.println("执行过滤操作");
      chain.doFilter(request, response);}
   @Override
   public void destroy() {
```

```
        System.out.println("过滤器销毁");    }
}
```

4.8.8　创建类 MyServletContextListener

创建类 MyServletContextListener，代码如例 4-34 所示。

【例 4-34】 创建类 MyServletContextListener 的代码示例。

```
package com.bookcode.servlet;
import javax.servlet.ServletContextEvent;
import javax.servlet.ServletContextListener;
import javax.servlet.annotation.WebListener;
@WebListener
public class MyServletContextListener implements ServletContextListener {
    @Override
    public void contextDestroyed(ServletContextEvent arg0) {
        System.out.println("ServletContex 销毁");    }
    @Override
    public void contextInitialized(ServletContextEvent arg0) {
        System.out.println("ServletContex 初始化");    }
}
```

4.8.9　创建类 MyHttpSessionListener

创建类 MyHttpSessionListener，代码如例 4-35 所示。

【例 4-35】 创建类 MyHttpSessionListener 的代码示例。

```
package com.bookcode.servlet;
import javax.servlet.annotation.WebListener;
import javax.servlet.http.HttpSessionEvent;
import javax.servlet.http.HttpSessionListener;
@WebListener
public class MyHttpSessionListener implements HttpSessionListener {
    @Override
    public void sessionCreated(HttpSessionEvent se) {
        System.out.println("Session 被创建");    }
    @Override
    public void sessionDestroyed(HttpSessionEvent se) {
        System.out.println("ServletContex 初始化");    }
}
```

4.8.10　运行程序 3

运行程序后，控制台中的输出结果如图 4-15 所示。再在浏览器中输入对任意一个页面的访问（如 http://localhost:8080/myServlet2），控制台中的输出结果如图 4-16 所示。

```
ServletContex初始化
过滤器初始化
```

图 4-15　程序运行后在控制台中的输出结果

```
执行过滤操作
```

图 4-16　在浏览器中输入对任意一个页面的访问后在控制台中的输出结果

4.8.11　创建类 MyInterceptor1

创建类 MyInterceptor1，代码如例 4-36 所示。

【例 4-36】 创建类 MyInterceptor1 的代码示例。

```
package com.bookcode.servlet;
import org.springframework.web.servlet.HandlerInterceptor;
import org.springframework.web.servlet.ModelAndView;
import javax.servlet.http.HttpServletRequest;
import javax.servlet.http.HttpServletResponse;
public class MyInterceptor1 implements HandlerInterceptor {
    @Override
  public boolean preHandle(HttpServletRequest request, HttpServletResponse
  response, Object handler)
            throws Exception {
        System.out.println(">>>MyInterceptor1>>>>>>>在请求处理之前进行调用
        (Controller 方法调用之前）");
                return true;//只有返回 true 才会继续向下执行，返回 false 取消当前请求
      }
    @Override
    public void postHandle(HttpServletRequest request, HttpServletResponse
    response, Object handler,
                        ModelAndView modelAndView) throws Exception {
        System.out.println(">>>MyInterceptor1>>>>>>>请求处理之后进行调用，但是
        在视图被渲染之前（Controller 方法调用之后）");
    }
    @Override
public void afterCompletion(HttpServletRequest request, HttpServletResponse
response, Object handler, Exception ex)
            throws Exception {
        System.out.println(">>>MyInterceptor1>>>>>>>在整个请求结束之后被调用，
```

```
        也就是在 DispatcherServlet 渲染了对应的视图之后执行（主要是用于进行资源清理
        工作）");
    }
}
```

4.8.12　创建类 MyInterceptor2

创建类 MyInterceptor2，代码如例 4-37 所示。

【例 4-37】 创建类 MyInterceptor2 的代码示例。

```
package com.bookcode.servlet;
import org.springframework.web.servlet.HandlerInterceptor;
import org.springframework.web.servlet.ModelAndView;
import javax.servlet.http.HttpServletRequest;
import javax.servlet.http.HttpServletResponse;
public class MyInterceptor2 implements HandlerInterceptor {
    @Override
  public boolean preHandle(HttpServletRequest request, HttpServletResponse
  response, Object handler)
            throws Exception {
        System.out.println(">>>MyInterceptor2>>>>>>>在请求处理之前进行调用
        (Controller 方法调用之前）");
                return true;//只有返回 true 才会继续向下执行，返回 false 取消当前请求
      }
    @Override
    public void postHandle(HttpServletRequest request, HttpServletResponse
    response, Object handler,
                         ModelAndView modelAndView) throws Exception {
        System.out.println(">>>MyInterceptor2>>>>>>>请求处理之后进行调用，但是
        在视图被渲染之前（Controller 方法调用之后）");
    }
    @Override
public void afterCompletion(HttpServletRequest request, HttpServletResponse
response, Object handler, Exception ex)
            throws Exception {
        System.out.println(">>>MyInterceptor2>>>>>>>在整个请求结束之后被调用，
        也就是在 DispatcherServlet 渲染了对应的视图之后执行（主要是用于进行资源清理
        工作）");
    }
}
```

4.8.13　创建类 MyWebAppConfigurer

创建类 MyWebAppConfigurer，代码如例 4-38 所示。

【例 4-38】 创建类 MyWebAppConfigurer 的代码示例。

```
package com.bookcode.config;
import com.bookcode.servlet.MyInterceptor1;
import com.bookcode.servlet.MyInterceptor2;
import org.springframework.context.annotation.Configuration;
import org.springframework.web.servlet.config.annotation.InterceptorRegistry;
import org.springframework.web.servlet.config.annotation.WebMvcConfigurerAdapter;
@Configuration
public class MyWebAppConfigurer extends WebMvcConfigurerAdapter {
    @Override
    public void addInterceptors(InterceptorRegistry registry) {
//多个拦截器组成一个拦截器链
//addPathPatterns 用于添加拦截规则
//excludePathPatterns 用户排除拦截
        registry.addInterceptor(new MyInterceptor1()).addPathPatterns("/**");
        registry.addInterceptor(new MyInterceptor2()).addPathPatterns("/**");
        super.addInterceptors(registry);
    }
}
```

4.8.14 运行程序 4

运行程序后，在浏览器中输入对任意一个页面的访问（如 http://localhost:8080/index），控制台中的结果如图 4-17 所示。

```
>>>MyInterceptor1>>>>>>>在请求处理之前进行调用（Controller方法调用之前）
>>>MyInterceptor2>>>>>>>在请求处理之前进行调用（Controller方法调用之前）
>>>MyInterceptor2>>>>>>>请求处理之后进行调用，但是在视图被渲染之前（Controller方法调用之后）
>>>MyInterceptor1>>>>>>>请求处理之后进行调用，但是在视图被渲染之前（Controller方法调用之后）
>>>MyInterceptor2>>>>>>>在整个请求结束之后被调用，也就是在DispatcherServlet 渲染了对应的视图之后执行（主要是用于进行资源清理工作）
>>>MyInterceptor1>>>>>>>在整个请求结束之后被调用，也就是在DispatcherServlet 渲染了对应的视图之后执行（主要是用于进行资源清理工作）
```

图 4-17 在浏览器中输入对任意一个页面的访问后在控制台中的结果

习题 4

简答题

1. 简述 Thymeleaf 的基础语法与用法。
2. 简述对 Freemarker 的理解。
3. 简述对 Ajax 的理解。
4. 简述对 RESTful 风格的理解。
5. 简述对 Bootstrap 的理解。

6．简述对 jQuery 的理解。

7．简述过滤器、监听器和拦截器的不同之处。

实验题

1．实现基于 Thymeleaf 的动态 Web 页面。

2．实现基于 Freemarker 的 Web 应用。

3．实现对 Ajax 的应用。

4．实现 RESTful 风格的 Web 应用。

5．实现带 Bootstrap 和 jQuery 的 Web 应用。

6．实现使用了 Servlet、过滤器、监听器和拦截器的 Web 应用。

第 5 章

Spring Boot 的数据库访问

本章结合实例介绍如何使用 JDBC 访问 H2 数据库、如何使用 Spring Data JPA 访问 H2 数据库、如何使用 Spring Data JPA 和 RESTful 访问 H2 数据库、如何使用 Spring Data JPA 访问 MySQL 数据库、如何访问 MongoDB 数据库、如何访问 Neo4j 数据库和访问数据库完整示例。

MongoDB 数据库、Neo4j 数据库和第 9 章将要介绍的 Redis 数据库都属于 NoSQL（Not Only SQL）数据库，NoSQL 数据库泛指非关系型的数据库。随着互联网 Web 2.0 网站（特别是超大规模和高并发的 SNS 类型 Web 2.0 纯动态网站）的兴起，传统的关系数据库暴露了很多难以克服的问题。于是，非关系型数据库得到了迅速的发展。

对于 NoSQL 数据库并没有一个明确的范围和定义，但是它们有一些相似特征。

（1）不需要事先定义数据模式和预定义表结构。

（2）NoSQL 数据库往往将数据划分后存储在各个本地服务器上。

（3）可以在系统运行的时候，动态增加或者删除结点。不需要停机维护，数据可以自动迁移。

（4）NoSQL 数据库中的复制，往往是基于日志的异步复制。这样，数据就可以尽快地写入到一个结点，而不会被网络传输引起迟延。其缺点是并不能总是保证数据一致性，出现故障时可能会丢失少量的数据。

（5）相对于事务严格的 ACID 特性，NoSQL 数据库保证的是 BASE 特性（最终一致性和软事务）。

NoSQL 数据库比较适用于数据模型比较简单、需要更强灵活性、对数据库性能要求较高、不需要高度的数据一致性等环境。NoSQL 数据库并没有一个统一的架构，两种 NoSQL 数据库之间的不同远远超过两种关系型数据库的不同。

5.1　使用 JDBC 访问 H2 数据库

本节介绍如何使用 Spring 的 JdbcTemplate 类来构建一个应用程序访问存储在关系型数据库（本节使用 H2 数据库）中的数据。

5.1.1　添加依赖

视频讲解

因为使用 JDBC 访问 H2 数据库，所以需要增加对 JDBC 和 H2 的依赖。因为要通过浏览器来访问 H2 数据库的控制台，所以需要有 Web 依赖。首先，在 pom.xml 文件中，在<dependencies>和</dependencies>之间添加 JDBC、H2 和 Web 依赖，代码如例 5-1 所示。

【例 5-1】 添加 JDBC、H2 和 Web 依赖的代码示例。

```
<dependency>
     <groupId>org.springframework.boot</groupId>
     <artifactId>spring-boot-starter-jdbc</artifactId>
</dependency>
<dependency>
     <groupId>com.h2database</groupId>
     <artifactId>h2</artifactId>
</dependency>
<dependency>
     <groupId>org.springframework.boot</groupId>
     <artifactId>spring-boot-starter-web</artifactId>
</dependency>
```

5.1.2　创建类 Customer

在包 com.bookcode.entity 中创建类 Customer，代码如例 5-2 所示。

【例 5-2】 创建类 Customer 的代码示例。

```
package com.bookcode.entity;
public class Customer {
   private long id;
   private String firstName;
   private String lastName;
   public long getId() {
       return id;
    }
    public void setId(long id) {
       this.id = id;
    }
```

```
    public String getFirstName() {
        return firstName;
    }
    public void setFirstName(String firstName) {
        this.firstName = firstName;
    }
    public String getLastName() {
        return lastName;
    }
    public void setLastName(String lastName) {
        this.lastName = lastName;
    }
    public Customer(long id, String firstName, String lastName) {
        this.id = id;
        this.firstName = firstName;
        this.lastName = lastName;
    }
    @Override
    public String toString() {
        return String.format("Customer[id=%d, firstName='%s', lastName='%s']",
                id, firstName, lastName);
    }
}
```

5.1.3 修改入口类

修改入口类，代码如例 5-3 所示。

【例 5-3】 修改入口类的代码示例。

```
package com.bookcode;
import com.bookcode.entity.Customer;
import org.slf4j.Logger;
import org.slf4j.LoggerFactory;
import org.springframework.beans.factory.annotation.Autowired;
import org.springframework.boot.CommandLineRunner;
import org.springframework.boot.SpringApplication;
import org.springframework.boot.autoconfigure.SpringBootApplication;
import org.springframework.jdbc.core.JdbcTemplate;
import java.util.Arrays;
import java.util.List;
import java.util.stream.Collectors;
@SpringBootApplication
```

```
public class DemoApplication implements CommandLineRunner{
    private static final Logger log = LoggerFactory.getLogger(DemoApplication.class);
    public static void main(String[] args) {
        SpringApplication.run(DemoApplication.class, args);    }
    @Autowired                              //注解@Autowired 完成自动装配的工作
     JdbcTemplate  jdbcTemplate;    //对数据库的所有操作将要借助于 JdbcTemplate
    @Override
    public void run(String... strings) throws Exception {
        log.info("Creating tables");//日志信息输出到控制台
        //创建表 customers
        jdbcTemplate.execute("DROP TABLE customers IF EXISTS");
        jdbcTemplate.execute("CREATE TABLE customers(" +
                "id SERIAL, first_name VARCHAR(255), last_name VARCHAR(255))");
        //将整个数组的数组拆分成 firstName、lastName 的数组
        List<Object[]> splitUpNames = Arrays.asList("John Woo", "Jeff Dean",
        "Josh Bloch",
                "Josh Long").stream()
                .map(name -> name.split(" "))
                .collect(Collectors.toList());
        //使用 Java 8 流以日志信息形式输出每个元组的列表到控制台
        splitUpNames.forEach(name -> log.info(String.format("Inserting customer
        record for %s %s",
                name[0], name[1])));
        //使用 JdbcTemplate 批处理更新数据
  jdbcTemplate.batchUpdate("INSERT INTO customers(first_name, last_name)
  VALUES (?,?)", splitUpNames);
        log.info("Querying for customer records where first_name = 'Josh':");
        //查询 first_name 为'Josh'的 customer 信息（id, first_name, last_name)
        jdbcTemplate.query(
  "SELECT id, first_name, last_name FROM customers WHERE first_name = ?",  new
  Object[] { "Josh" },
                (rs, rowNum) -> new Customer(rs.getLong("id"), rs.getString
                ("first_name"), rs.getString("last_name")            )
        ).forEach(customer -> log.info(customer.toString()));
    }
}
```

5.1.4　修改配置文件 application.properties

为了更好地观察对 H2 的访问情况，通过修改配置文件 application.properties 使 H2 数据库的控制台可用，代码如例 5-4 所示。

【例 5-4】 修改配置文件 application.properties 的代码示例。

```
#为了更好地观察对 H2 的访问情况，设置使 H2 控制台可用
spring.h2.console.enabled=true
```

5.1.5 运行程序

运行程序，数据库处理信息在控制台中输出，结果如图 5-1 所示。

```
o.s.b.w.embedded.tomcat.TomcatWebServer   : Tomcat started on port(s): 8080 (http) with context path ''
c.b.SpringbootHelloworldsApplication      : Started SpringbootHelloworldsApplication in 3.04 seconds (JVM running for 4.506)
c.b.SpringbootHelloworldsApplication      : Creating tables
com.zaxxer.hikari.HikariDataSource        : HikariPool-1 - Starting...
com.zaxxer.hikari.HikariDataSource        : HikariPool-1 - Start completed.
c.b.SpringbootHelloworldsApplication      : Inserting customer record for John Woo
c.b.SpringbootHelloworldsApplication      : Inserting customer record for Jeff Dean
c.b.SpringbootHelloworldsApplication      : Inserting customer record for Josh Bloch
c.b.SpringbootHelloworldsApplication      : Inserting customer record for Josh Long
c.b.SpringbootHelloworldsApplication      : Querying for customer records where first_name = 'Josh':
c.b.SpringbootHelloworldsApplication      : Customer[id=3, firstName='Josh', lastName='Bloch']
c.b.SpringbootHelloworldsApplication      : Customer[id=4, firstName='Josh', lastName='Long']
o.a.c.c.C.[Tomcat].[localhost].[/]        : Initializing Spring FrameworkServlet 'dispatcherServlet'
o.s.web.servlet.DispatcherServlet         : FrameworkServlet 'dispatcherServlet': initialization started
o.s.web.servlet.DispatcherServlet         : FrameworkServlet 'dispatcherServlet': initialization completed in 33 ms
```

图 5-1 数据库处理信息在控制台中的输出结果

为了更好地观察对数据库的访问情况，可以通过在浏览器中输入 localhost:8080/h2-console 来访问 H2 数据库的控制台。H2 控制台启动界面如图 5-2 所示。单击“连接”按钮，可以看到 H2 数据库信息，如图 5-3 所示。图 5-3 还显示了对表 customers 执行 SELECT * FROM CUSTOMERS 操作后显示 customers 全体记录的情况，这和图 5-1 中输出信息吻合。

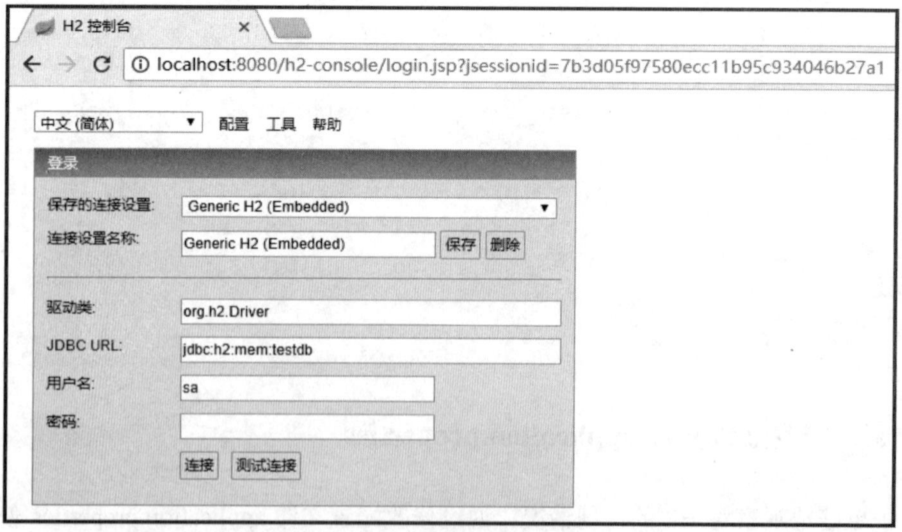

图 5-2 H2 控制台启动界面

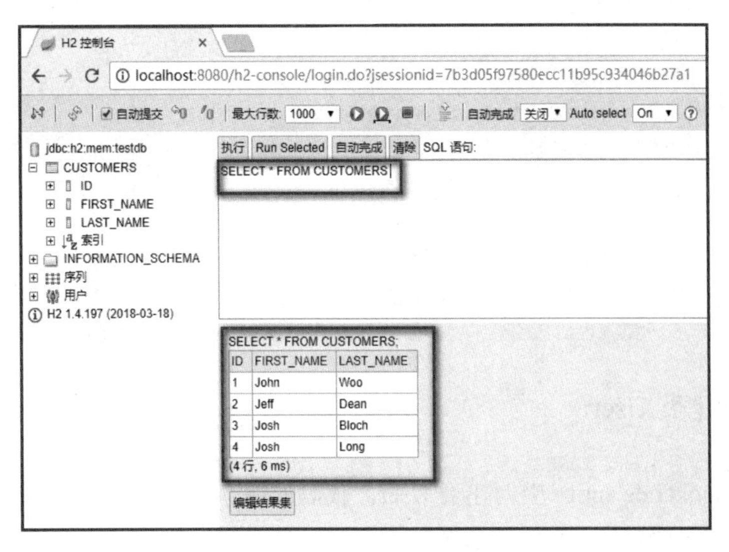

图 5-3　H2 数据库信息

5.2　使用 Spring Data JPA 访问 H2 数据库

视频讲解

JPA 是 Java 持久层 API（Java Persistence API）的简称。JPA 是 JCP 组织发布的 Java EE 标准之一，任何声称符合 JPA 标准的框架都遵循同样的架构，提供相同的访问 API。这保证了基于 JPA 开发的企业应用能够很容易在不同的 JPA 框架下运行。

引入新的 JPA 规范有两方面的原因：一方面，是为了简化现有 Java EE 和 Java SE 应用开发工作；另一方面，是为了整合对象关系映射（ORM）技术。JPA 框架中支持大数据集、事务、并发等容器级事务，这使得 JPA 超越了简单持久化框架的局限，在企业应用中发挥更大的作用。JPA 的一个主要目标就是提供更加简单的编程模型。在 JPA 框架下创建实体很简单，只需要使用@Entity 注解进行标注；JPA 的框架和接口也非常简单，开发者可以很容易地掌握。JPA 的设计是基于非侵入式原则的，因此可以很容易地和其他框架或者容器集成。

JPA 是一套标准、规范，不是产品；Hibernate、TopLink、JDO、Spring Data JPA 是具体的产品。为了说明如何使用 Spring Data JPA 访问数据库，本节介绍使用 Spring Data JPA（后面章节添加依赖时简称 Data JPA）来访问 H2 数据库。

5.2.1　添加依赖

首先，在 pom.xml 文件中<dependencies>和</dependencies>之间添加 Data JPA 和 H2、Web 依赖，代码如例 5-5 所示。

【例 5-5】 添加 Data JPA 和 H2、Web 依赖的代码示例。

```
<dependency>
    <groupId>org.springframework.boot</groupId>
    <artifactId>spring-boot-starter-data-jpa</artifactId>
```

```
</dependency>
<dependency>
      <groupId>com.h2database</groupId>
      <artifactId>h2</artifactId>
</dependency>
<dependency>
      <groupId>org.springframework.boot</groupId>
      <artifactId>spring-boot-starter-web</artifactId>
</dependency>
```

5.2.2 创建类 User

在包 com.bookcode.entity 中创建类 User，代码如例 5-6 所示。

【例 5-6】 创建类 User 的代码示例。

```
package com.bookcode.entity;
import javax.persistence.Entity;
import javax.persistence.GeneratedValue;
import javax.persistence.GenerationType;
import javax.persistence.Id;
@Entity                                   //注解@Entity 说明是数据实体类
public class User {
    @Id                                   //注解@Id 指明将属性 id 映射为数据库的主键
    @GeneratedValue(strategy=GenerationType.IDENTITY)//采用自增策略增加 id 值
    private Long id;
    private String firstName;
    private String lastName;
    protected User() {}
    public User(String firstName, String lastName) {
        this.firstName = firstName;
        this.lastName = lastName;
    }
    @Override
    public String toString() {
        return String.format(
                "User[id=%d, firstName='%s', lastName='%s']",
                id, firstName, lastName);
    }
}
```

5.2.3 创建接口 UserRepository

在包 com.bookcode.dao 中创建接口 UserRepository，代码如例 5-7 所示。UserRepository 接口扩展了 JpaRepository 接口；JpaRepository 封装了对数据库的访问方法，其泛型参数依次为实体类名和实体中主键 id 的类型。

【例 5-7】 创建接口 UserRepository 的代码示例。

```
package com.bookcode.dao;
import com.bookcode.entity.User;
import org.springframework.data.jpa.repository.JpaRepository;
public interface UserRepository extends JpaRepository<User,Long> {
}
```

5.2.4 修改入口类

修改入口类，代码如例 5-8 所示。

【例 5-8】 修改入口类的代码示例。

```
package com.bookcode;
import com.bookcode.dao.UserRepository;
import com.bookcode.entity.User;
import org.slf4j.Logger;
import org.slf4j.LoggerFactory;
import org.springframework.boot.CommandLineRunner;
import org.springframework.boot.SpringApplication;
import org.springframework.boot.autoconfigure.SpringBootApplication;
import org.springframework.context.annotation.Bean;
@SpringBootApplication
public class DemoApplication  {
    private static final Logger log = LoggerFactory.getLogger(DemoApplication.class);
    public static void main(String[] args) {
        SpringApplication.run(DemoApplication.class, args);    }
    @Bean
    public CommandLineRunner demo(UserRepository repository) {
        return (args) -> {
            //存储 5 条用户记录到数据库
            repository.save(new User("Jack", "Bauer"));
            repository.save(new User("Chloe", "O'Brian"));
            repository.save(new User("Kim", "Bauer"));
            repository.save(new User("David", "Palmer"));
            repository.save(new User("Michelle", "Dessler"));
            //将所有 user 信息以日志形式输出到控制台
            log.info("Users found with findAll():");
            log.info("-------------------------------");
            for (Object user : repository.findAll()) {
                log.info(user.toString());          }
            //获取 id=1 的 user 信息，并以日志形式在控制台输出
            repository.findById(1L)
                    .ifPresent(User -> {
                        log.info("User found with findById(1L):");
```

```
                    log.info("--------------------------------");
                    log.info(User.toString());
                });
        };
    }
}
```

5.2.5 修改配置文件 application.properties

为了更好地观察对 H2 数据库的访问情况，通过修改配置文件 application.properties 使 H2 数据库的控制台可用，代码如例 5-4 所示。

5.2.6 运行程序

运行程序，数据库处理信息在控制台中的输出结果如图 5-4 所示。在 H2 数据库的控制台中看到的 H2 数据库信息如图 5-5 所示。

```
o.s.b.w.embedded.tomcat.TomcatWebServer  : Tomcat started on port(s): 8080 (http) with context path ''
c.b.SpringbootHelloworldsApplication     : Started SpringbootHelloworldsApplication in 6.63 seconds (JVM running for 8.497)
c.b.SpringbootHelloworldsApplication     : Users found with findAll():
c.b.SpringbootHelloworldsApplication     : -------------------------------
o.h.h.i.QueryTranslatorFactoryInitiator  : HHH000397: Using ASTQueryTranslatorFactory
c.b.SpringbootHelloworldsApplication     : User[id=1, firstName='Jack', lastName='Bauer']
c.b.SpringbootHelloworldsApplication     : User[id=2, firstName='Chloe', lastName='O'Brian']
c.b.SpringbootHelloworldsApplication     : User[id=3, firstName='Kim', lastName='Bauer']
c.b.SpringbootHelloworldsApplication     : User[id=4, firstName='David', lastName='Palmer']
c.b.SpringbootHelloworldsApplication     : User[id=5, firstName='Michelle', lastName='Dessler']
c.b.SpringbootHelloworldsApplication     : User found with findById(1L):
c.b.SpringbootHelloworldsApplication     : --------------------------------
c.b.SpringbootHelloworldsApplication     : User[id=1, firstName='Jack', lastName='Bauer']
```

图 5-4 数据库处理信息在控制台中的输出结果

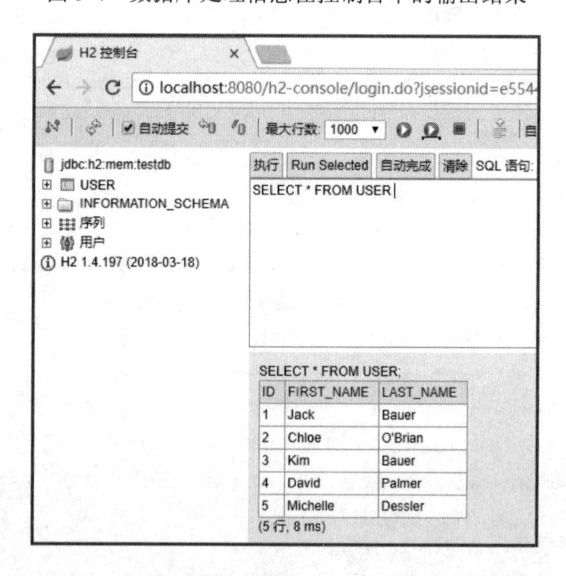

图 5-5 在 H2 数据库的控制台中看到的 H2 数据库信息

5.2.7　程序扩展

在包 com.bookcode.dao 中创建接口 UserCrudRepository，代码如例 5-9 所示。UserCrudRepository 接口继承了 CrudRepository 接口，并增加了一个访问数据库的方法。

【例 5-9】 创建接口 UserCrudRepository 的代码示例。

```
package com.bookcode.dao;
import com.bookcode.entity.User;
import org.springframework.data.repository.CrudRepository;
import java.util.List;
public interface UserCrudRepository extends CrudRepository<User, Long> {
    List<User> findByLastName(String lastName);
}
```

修改入口类，代码如例 5-10 所示。

【例 5-10】 修改入口类的代码示例。

```
package com.bookcode;
import com.bookcode.entity.User;
import com.bookcode.dao.UserCrudRepository;
import org.slf4j.Logger;
import org.slf4j.LoggerFactory;
import org.springframework.boot.CommandLineRunner;
import org.springframework.boot.SpringApplication;
import org.springframework.boot.autoconfigure.SpringBootApplication;
import org.springframework.context.annotation.Bean;
@SpringBootApplication
public class DemoApplication  {
    private static final Logger log = LoggerFactory.getLogger(DemoApplication.class);
    public static void main(String[] args) {
        SpringApplication.run(DemoApplication.class, args);    }
    @Bean
    public CommandLineRunner demo(UserCrudRepository repository) {
        return (args) -> {
            repository.save(new User("Jack", "Bauer"));
            repository.save(new User("Chloe", "O'Brian"));
            repository.save(new User("Kim", "Bauer"));
            repository.save(new User("David", "Palmer"));
            repository.save(new User("Michelle", "Dessler"));
            log.info("");                              //在控制台中输出空行
            //查找 lastName 为'Bauer'的所有 user，并以日志形式输出到控制台
            log.info("Customer found with findByLastName('Bauer'):");
            log.info("--------------------------------------------");
```

```
            repository.findByLastName("Bauer").forEach(bauer -> {
                log.info(bauer.toString());
            });
        };
    }
}
```

运行程序，处理数据库的信息在控制台中的输出结果如图 5-6 所示。

```
o.s.b.w.embedded.tomcat.TomcatWebServer  : Tomcat started on port(s): 8080 (http) with context path ''
c.b.SpringbootHelloworldsApplication     : Started SpringbootHelloworldsApplication in 5.06 seconds (JVM running for 7.033)
c.b.SpringbootHelloworldsApplication     :
c.b.SpringbootHelloworldsApplication     : Customer found with findByLastName('Bauer'):
c.b.SpringbootHelloworldsApplication     : --------------------------------------------
o.h.h.i.QueryTranslatorFactoryInitiator  : HHH000397: Using ASTQueryTranslatorFactory
c.b.SpringbootHelloworldsApplication     : User[id=1, firstName='Jack', lastName='Bauer']
c.b.SpringbootHelloworldsApplication     : User[id=3, firstName='Kim', lastName='Bauer']
```

图 5-6　数据库处理信息在控制台中的输出结果（lastName 为 Bauer 的所有 user 信息）

5.3　使用 Spring Data JPA 和 RESTful 访问 H2 数据库

5.3.1　添加依赖

视频讲解

在 pom.xml 文件中<dependencies>和</dependencies>之间添加 Data JPA、REST、H2 和 Web 依赖，代码如例 5-11 所示。

【例 5-11】 添加 Data JPA、REST、H2 和 Web 依赖的代码示例。

```
<dependency>
    <groupId>org.springframework.boot</groupId>
    <artifactId>spring-boot-starter-data-jpa</artifactId>
</dependency>
<dependency>
    <groupId>org.springframework.data</groupId>
    <artifactId>spring-data-rest-core</artifactId>
</dependency>
<dependency>
    <groupId>com.h2database</groupId>
    <artifactId>h2</artifactId>
</dependency>
<dependency>
    <groupId>org.springframework.boot</groupId>
    <artifactId>spring-boot-starter-web</artifactId>
</dependency>
```

5.3.2　创建类 Person

在包 com.bookcode.entity 中创建类 Person，代码如例 5-12 所示。

【例 5-12】 创建类 Person 的代码示例。

```
package com.bookcode.entity;
import javax.persistence.Entity;
import javax.persistence.GeneratedValue;
import javax.persistence.GenerationType;
import javax.persistence.Id;
@Entity
public class Person {
    @Id
    @GeneratedValue(strategy = GenerationType.AUTO)
    private Long id;
    private String firstName;
    private String lastName;
    public String getFirstName() {
        return firstName;
    }
    public void setFirstName(String firstName) {
        this.firstName = firstName;
    }
    public String getLastName() {
        return lastName;
    }
    public void setLastName(String lastName) {
        this.lastName = lastName;
    }
}
```

5.3.3 创建接口 PersonRepository

在包 com.bookcode.dao 中创建接口 PersonRepository，代码如例 5-13 所示。PersonRepository 接口继承了 PagingAndSortingRepository 接口，并增加了一个访问数据库的方法。

【例 5-13】 创建接口 PersonRepository 的代码示例。

```
package com.bookcode.dao;
package com.bookcode;
import java.util.List;
import org.springframework.data.repository.PagingAndSortingRepository;
import org.springframework.data.repository.query.Param;
import org.springframework.data.rest.core.annotation.RepositoryRestResource;
@RepositoryRestResource(collectionResourceRel = "people", path = "people")
//访问路径
public interface PersonRepository extends PagingAndSortingRepository
<Person, Long> {
    List<Person> findByLastName(@Param("name") String name);
}
```

5.3.4 修改配置文件 application.properties

为了更好地观察对 H2 数据库的访问情况，通过修改配置文件 application.properties 使 H2 数据库控制台可用，代码如例 5-4 所示。

5.3.5 启动程序并进行 REST 服务测试

测试用到的工具是 Postman。Postman 是一个支持 REST 的客户端，可以用它来测试资源。它可以被安装成 Chrome 插件，或者安装成独立应用程序。本书采用后一种方式。Postman 安装过程比较简单，本书不做介绍。

启动程序后可以利用 Postman 进行 REST 服务测试。当需要查询所有 people 时，先在 Postman 的 URL 处输入 http://localhost:8080/people，再选择 GET 方法，结果如图 5-7 所示。当需要增加一条记录（以增加“王三”的信息为例）时，在 Postman 的 URL 处输入 http://localhost:8080/people，并选择 POST 方法和 JOSN(application/json)格式，再以 JSON 格式输入“王三”的 firstName（“王”）和 lastName（“三”），结果如图 5-8 所示。搜索 lastName 为“三”的 people 信息时，先在 Postman 的 URL 处输入 http://localhost:8080/people/search/findByLastName?name=三，再选择 GET 方法，结果如图 5-9 所示。对 H2 数据库处理的结果在 H2 数据库控制台中的输出如图 5-10 所示。

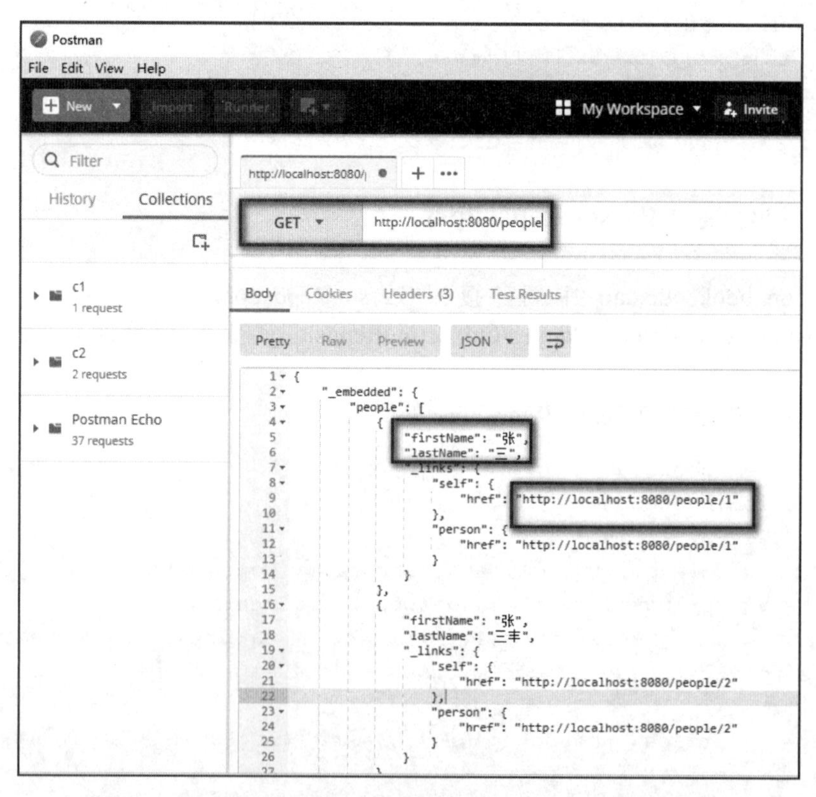

图 5-7 在 Postman 的 URL 处输入 http://localhost:8080/people 查询全体 people 信息的结果

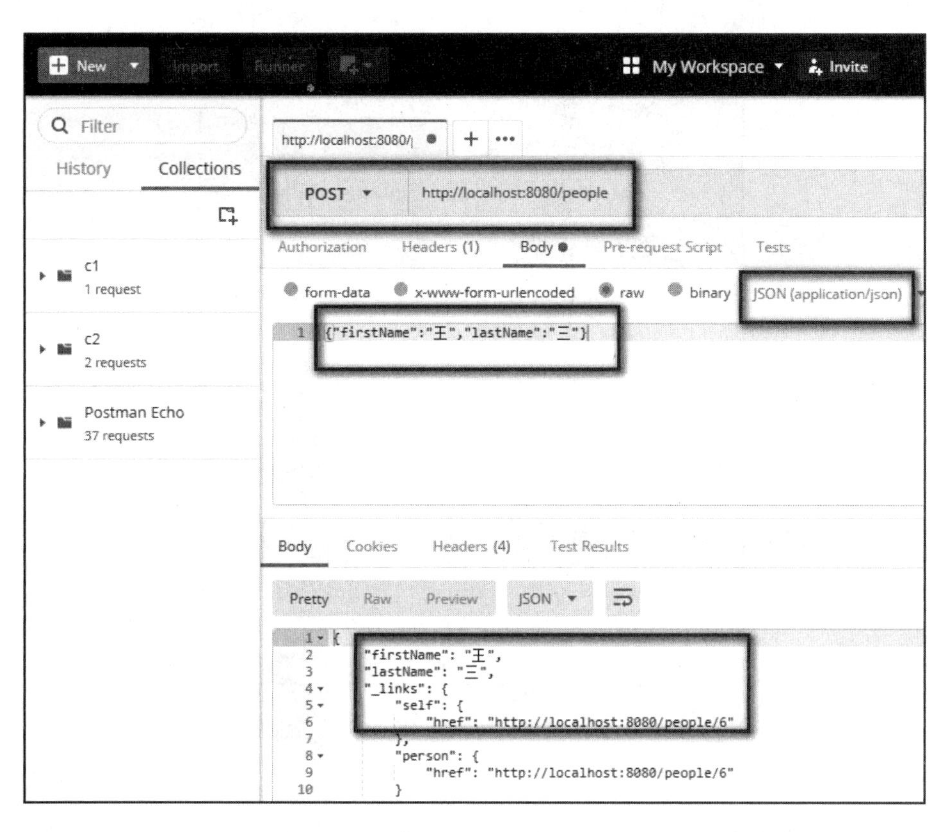

图 5-8　增加一条记录（“王三”的信息）的结果

```
"people": [
    {
        "firstName": "张",
        "lastName": "三",
        "_links": {
            "self": {
                "href": "http://localhost:8080/people/1"
            },
            "person": {
                "href": "http://localhost:8080/people/1"
            }
        }
    },
    {
        "firstName": "王",
        "lastName": "三",
        "_links": {
            "self": {
                "href": "http://localhost:8080/people/6"
            },
            "person": {
                "href": "http://localhost:8080/people/6"
            }
```

图 5-9　搜索 lastName 为“三”的结果

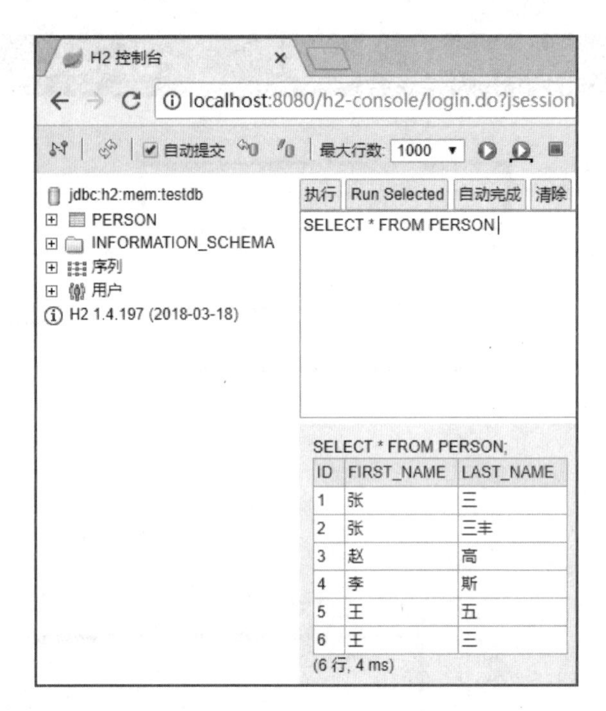

图 5-10　对 H2 数据库处理的结果在 H2 数据库控制台中的输出

5.4　使用 Spring Data JPA 访问 MySQL 数据库

视频讲解

本节介绍如何使用 Spring Data JPA 访问 MySQL 数据库。首先，需要安装 MySQL 数据库，在安装 MySQL 数据库时保留访问端口（3306）、用户名、密码信息，在访问数据库时要用到这些信息。MySQL 数据库安装过程比较简单，本书不做介绍。

5.4.1　添加依赖

首先，在 pom.xml 文件中<dependencies>和</dependencies>之间添加 Data JPA 依赖和 MySQL 驱动依赖，代码如例 5-14 所示。

【例 5-14】 添加 Data JPA 依赖和 MySQL 驱动依赖的代码示例。

```
<dependency>
     <groupId>org.springframework.boot</groupId>
     <artifactId>spring-boot-starter-data-jpa</artifactId>
</dependency>
<!--使用 MySQL 的 Connector/J 驱动-->
<dependency>
      <groupId>mysql</groupId>
      <artifactId>mysql-connector-java</artifactId>
</dependency>
```

5.4.2　创建类 User 和接口 UserRepository

在包 com.bookcode.entity 中创建类 User，代码如例 5-15 所示。在包 com.bookcode.dao 中创建接口 UserRepository，代码如例 5-7 所示。

【例 5-15】 创建类 User 的代码示例。

```
package com.bookcode.entity;
import javax.persistence.*;
@Entity
@Table(name="user")                         //增加此行语句来指定所映射的表名
public class User {
    @Id
    @GeneratedValue(strategy=GenerationType.IDENTITY)
    private Long id;
    private String firstName;
    private String lastName;
    protected User() {}
    public User(String firstName, String lastName) {
        this.firstName = firstName;
        this.lastName = lastName;
     }
    @Override
    public String toString() {
        return String.format(
                "User[id=%d, firstName='%s', lastName='%s']",
                id, firstName, lastName);
    }
}
```

5.4.3　修改配置文件和入口类

在配置文件 application.properties 中增加 MySQL 数据源配置信息，代码如例 5-16 所示。

【例 5-16】 增加 MySQL 数据源配置信息的代码示例。

```
#MySQL 数据源基本配置信息
#指定数据库
spring.datasource.url=jdbc:mysql://localhost:3306/mytest
#驱动程序
spring.datasource.driver-class-name=com.mysql.jdbc.Driver
spring.datasource.username=root
spring.datasource.password=sa
#更新时修改数据库
spring.jpa.hibernate.ddl-auto=update
```

修改入口类，代码如例 5-8 所示。

5.4.4 运行程序

运行程序后，控制台的输出信息与图 5-6 相同。同时，MySQL 的数据库 mytest 中自动生成了表 user，并且插入了 5 条 user 记录，对 MySQL 数据库的处理结果在工具 Navicat for MySQL 中的输出情况如图 5-11 所示。工具 Navicat for MySQL 的安装过程比较简单，本书不做介绍。

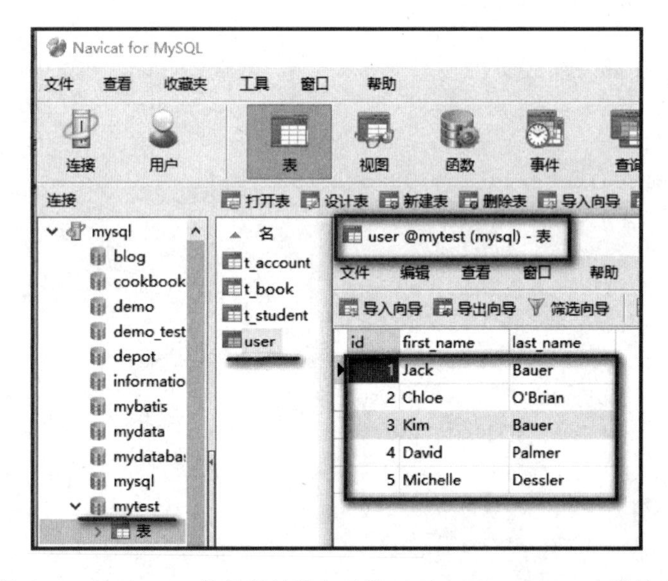

图 5-11　对 MySQL 的处理结果在工具 Navicat for MySQL 中的输出

5.4.5 程序扩展

在包 com.bookcode.controller 中创建类 UserController，其代码如例 5-17 所示。在包 com.bookcode.dao 中增加接口 UserCrudRepository，其代码如例 5-9 所示。

【例 5-17】 创建类 UserController 的代码示例。

```
package com.bookcode.controller;
import com.bookcode.dao.UserCrudRepository;
import com.bookcode.entity.User;
import org.springframework.beans.factory.annotation.Autowired;
import org.springframework.stereotype.Controller;
import org.springframework.web.bind.annotation.*;
import java.util.List;
@Controller                          //控制器
@RequestMapping(path="/demo")    //UserController 类中地址是相对地址，在/demo 后添加
public class UserController {
    @Autowired
```

```
    private UserCrudRepository userCrudRepository;
    @GetMapping(path="/add")     //相对地址，相当于/demo/add
    @ResponseBody                //@ResponseBody 表明返回是字符串而不是视图名
public  String addNewUser (@RequestParam String firstname , @RequestParam
String lastname) {
        //@RequestParam 表示传入 User 构造器中的参数
        User user = new User(firstname,lastname);
        userCrudRepository.save(user);
        return "Saved";
    }
    @GetMapping(path="/finduser/{lastname}") //根据 lastname 查找返回 user 信息
    @ResponseBody
                        //@PathVariable 表示参数 lastname
    public  String finduser  (@PathVariable("lastname")  String lastname){
        List<User> userList = userCrudRepository.findByLastName(lastname);
        String users ="  ";
        for(User user:userList) {users += user.toString() + "      ";};
        return users;
    }
}
```

运行程序后，为了增加 firstname 为 li 且 lastname 为 si 的 user，在浏览器中输入 localhost:8080/demo/add?firstname=li&lastname=si，结果如图 5-12 所示；数据库的 user 表中新增记录在工具 Navicat for MySQL 中的结果如图 5-13 所示。为了查找 lastname 为 Bauer 的所有 user 信息，需要在浏览器中输入 localhost:8080/demo/finduser/Bauer，结果如图 5-14 所示。

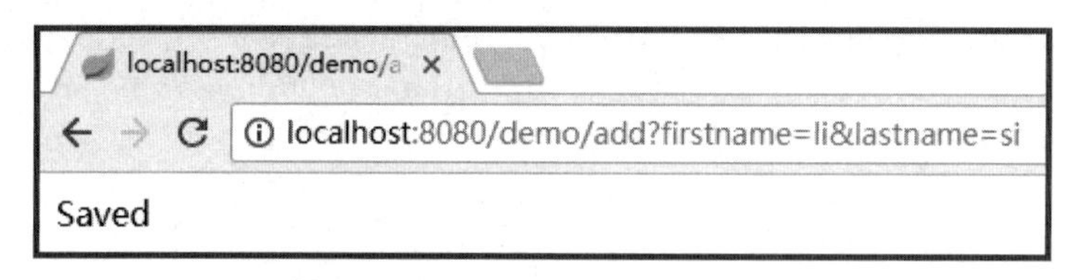

图 5-12　增加 user（firstname 为 li 且 lastname 为 si）结果

图 5-13　user 表新增记录在工具 Navicat for MySQL 中的结果

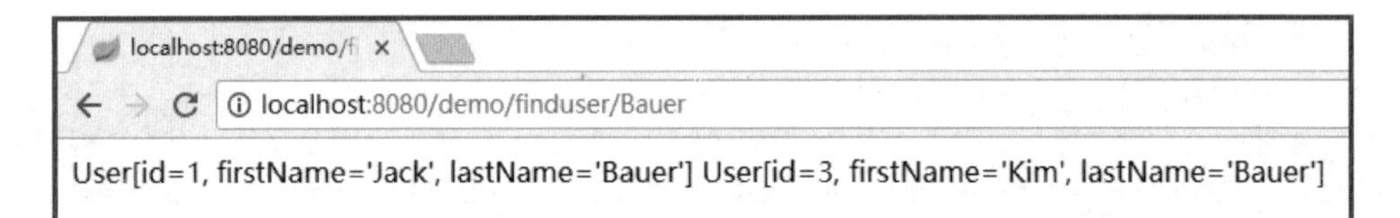

图 5-14 在浏览器中输入 localhost:8080/demo/finduser/Bauer
（查找 lastname 为 Bauer 的 user）的结果

5.5 访问 MongoDB 数据库

视频讲解

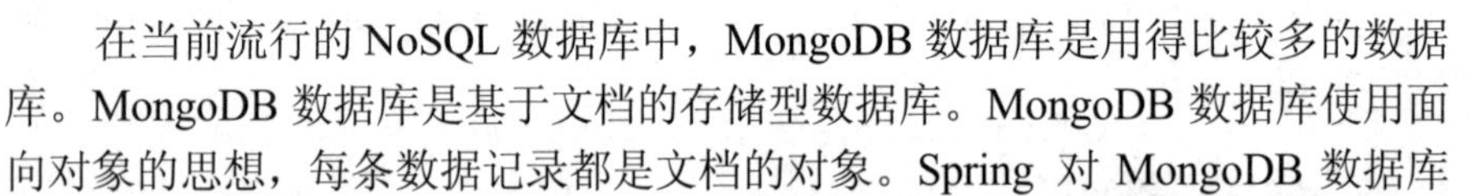

在当前流行的 NoSQL 数据库中，MongoDB 数据库是用得比较多的数据库。MongoDB 数据库是基于文档的存储型数据库。MongoDB 数据库使用面向对象的思想，每条数据记录都是文档的对象。Spring 对 MongoDB 数据库的支持主要是通过 Spring Data MongoDB 来实现的。由于 Spring Boot 对 MongoDB 数据库有专门的支持，Spring Boot 在访问 MongoDB 数据库时不需要进行配置。使用 MongoDB 数据库前需要先安装 MongoDB 数据库，MongoDB 数据库有 Window、Linux 等系统的安装包，安装过程比较简单，本书不做介绍。

5.5.1 添加依赖

在 pom.xml 文件中<dependencies>和</dependencies>之间添加 MongoDB、Web 等依赖，代码如例 5-18 所示。

【例 5-18】 添加 MongoDB、Web 等依赖的代码示例。

```
<dependency>
     <groupId>org.springframework.boot</groupId>
     <artifactId>spring-boot-starter-data-mongodb</artifactId>
</dependency>
<dependency>
       <groupId>javax.persistence</groupId>
       <artifactId>persistence-api</artifactId>
       <version>1.0</version>
</dependency>
<dependency>
       <groupId>org.springframework.boot</groupId>
       <artifactId>spring-boot-starter-web</artifactId>
</dependency>
```

5.5.2 创建类 Person

在包 com.bookcode.entity 中创建类 Person，代码如例 5-19 所示。

【例 5-19】 创建类 Person 的代码示例。

```
package com.bookcode.entity;
```

```
import javax.persistence.Id;
public class Person {
    @Id
    public String id;
    public String firstName;
    public String lastName;
    public Person() {}
    public Person(String firstName, String lastName) {
        this.firstName = firstName;
        this.lastName = lastName;
    }
    @Override
    public String toString() {
        return String.format(
                "Person[id=%s, firstName='%s', lastName='%s']",
                id, firstName, lastName);
    }
    public String getFirstName() {
        return firstName;
    }
    public void setFirstName(String firstName) {
        this.firstName = firstName;
    }
    public String getLastName() {
        return lastName;
    }
    public void setLastName(String lastName) {
        this.lastName = lastName;
    }
}
```

5.5.3　创建接口 PersonRepository

在包 com.bookcode.dao 中创建接口 PersonRepository，代码如例 5-20 所示。PersonRepository 接口继承并扩展了 MongoRepository 接口。MongoRepository 封装了对数据库 MongoDB 的访问方法，其泛型参数依次为实体类名和实体中主键 id 的类型。

【例 5-20】 创建接口 PersonRepository 的代码示例。

```
package com.bookcode.dao;
import com.bookcode.entity.Person;
import org.springframework.data.mongodb.repository.MongoRepository;
import java.util.List;
public interface PersonRepository extends MongoRepository<Person, String> {
    public Person findByFirstName(String firstName);
```

```
    public List<Person > findByLastName(String lastName);
}
```

5.5.4 修改入口类

修改入口类，代码如例 5-21 所示。

【例 5-21】 修改入口类的代码示例。

```
package com.bookcode;
import com.bookcode.dao.PersonRepository;
import com.bookcode.entity.Person;
import org.springframework.beans.factory.annotation.Autowired;
import org.springframework.boot.CommandLineRunner;
import org.springframework.boot.SpringApplication;
import org.springframework.boot.autoconfigure.SpringBootApplication;
@SpringBootApplication
public class DemoApplication implements CommandLineRunner {
        @Autowired
        private PersonRepository repository;
        public static void main (String[]args){
        SpringApplication.run(DemoApplication.class, args);
    }
        @Override
        public void run (String...args) throws Exception {
        repository.deleteAll();                        //辅助功能，先删除已有记录
        repository.save(new Person("Alice", "Smith")); //存储 Person 信息
        repository.save(new Person("Bob", "Smith"));    //存储 Person 信息
        //将所有 Person 信息输出
        System.out.println("Persons found with findAll():");
        System.out.println("-------------------------------");
        for (Person person : repository.findAll()) {
            System.out.println(person);
        }
        System.out.println();
        //找 firstName 为'Alice'的 Person，并将其信息输出到控制台
        System.out.println("Person found with findByFirstName('Alice'):");
        System.out.println("--------------------------------");
        System.out.println(repository.findByFirstName("Alice"));
        //找 lastName 为 Smith 的 Person，并将其信息输出到控制台
        System.out.println("Persons found with findByLastName('Smith'):");
        System.out.println("--------------------------------");
        for (Person Person : repository.findByLastName("Smith")) {
            System.out.println(Person);
```

```
        }
    }
}
```

5.5.5　运行程序

先在 Window 命令处理程序 CMD 中输入命令，启动 MongoDB 数据库，代码如例 5-22 所示。代码中路径和 MongoDB 数据库的设置有关。

【例 5-22】 启动 MongoDB 数据库的代码示例。

```
mongod --dbpath "D:\mongodb\data\db"
```

启动 MongoDB 数据库后，运行程序，控制台中的输出如图 5-15 所示。与此同时，MongoDB 数据库中的数据库 test 中自动生成了 2 条记录，在可视化工具 Robo 3T 中的显示情况如图 5-16 所示。可视化工具 Robo 3T 的安装过程比较简单，本书不做介绍。

```
Persons found with findAll():
-------------------------------
Person[id=5b0a2f549513f31e6cb493d3, firstName='Alice', lastName='Smith']
Person[id=5b0a2f549513f31e6cb493d4, firstName='Bob', lastName='Smith']

Person found with findByFirstName('Alice'):
--------------------------------
Person[id=5b0a2f549513f31e6cb493d3, firstName='Alice', lastName='Smith']
Persons found with findByLastName('Smith'):
--------------------------------
Person[id=5b0a2f549513f31e6cb493d3, firstName='Alice', lastName='Smith']
Person[id=5b0a2f549513f31e6cb493d4, firstName='Bob', lastName='Smith']
```

图 5-15　控制台中的输出

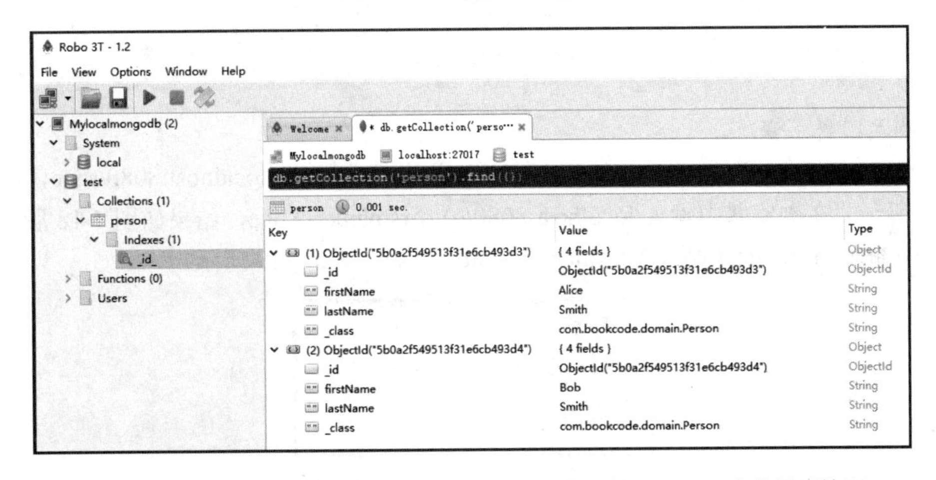

图 5-16　MongoDB 数据库增加 2 条记录在可视化工具 Robo 3T 中的显示情况

5.5.6 程序扩展

在包 com.bookcode.controller 中创建类 PersonController，代码如例 5-23 所示。

【例 5-23】 创建类 PersonController 的代码示例。

```
package com.bookcode.controller;
import com.bookcode.dao.PersonRepository;
import com.bookcode.entity.Person;
import org.springframework.beans.factory.annotation.Autowired;
import org.springframework.web.bind.annotation.RequestMapping;
import org.springframework.web.bind.annotation.RestController;
import java.util.List;
@RestController
public class PersonController {
    @Autowired
    private PersonRepository personRepository;
    @RequestMapping("/save")
    public String  save() {
        Person person = new Person("Li","wl");
        personRepository.save(person);
        return "Save OK!";    }
    @RequestMapping("/q1")
    public Person q1(String firstname) {
        Person person = personRepository.findByFirstName(firstname);
        return personRepository.save(person);    }
    @RequestMapping("/q2")
    public String q2(String lastname) {
        List<Person> listperson = personRepository.findByLastName(lastname);
        String persons ="  ";
        for(Person person:listperson) {persons += person + "      ";};
        return persons;          }
}
```

保持入口类不变。

启动 MongoDB 数据库后，运行程序。在浏览器中输入 localhost:8080/save，结果如图 5-17 所示。在浏览器中输入 localhost:8080/q1?firstname=Alice，结果如图 5-18 所示。在浏览器中输入 localhost:8080/q2?lastname=Smith，结果如图 5-19 所示。

图 5-17 在浏览器中输入 localhost:8080/save（成功增加 Person）的结果

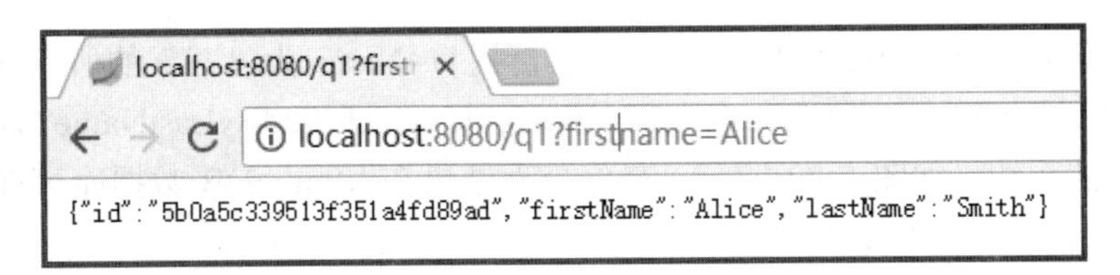

图 5-18　浏览器中输入 localhost:8080/q1?firstname=Alice
（查询 firstname 为 Alice 的 Person）的结果

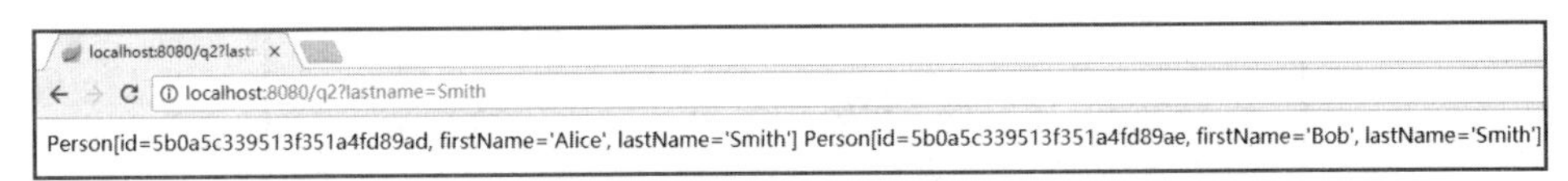

图 5-19　浏览器中输入 localhost:8080/q2?lastname=Smith
（返回 lastname 为 Smith 的 Person）的结果

5.5.7　使用 REST 方法访问 MongoDB

在 pom.xml 文件中<dependencies>和</dependencies>之间添加 REST 依赖，代码如例 5-24 所示。

【例 5-24】 添加 REST 依赖的代码示例。

```
<dependency>
        <groupId>org.springframework.boot</groupId>
        <artifactId>spring-boot-starter-data-rest</artifactId>
</dependency>
```

修改接口 PersonRepository，代码如例 5-25 所示。

【例 5-25】 修改接口 PersonRepository 的代码示例。

```
package com.bookcode.dao;
import com.bookcode.entity.Person;
import org.springframework.data.mongodb.repository.MongoRepository;
import org.springframework.data.repository.query.Param;
import org.springframework.data.rest.core.annotation.RepositoryRestResource;
import java.util.List;
@RepositoryRestResource(collectionResourceRel = "people", path = "people")
//路径
public interface PersonRepository extends MongoRepository<Person, String> {
    public Person findByFirstName(String firstName);
    public List<Person > findByLastName(@Param("name") String lastName);
}
```

启动 MongoDB 数据库，运行程序。为了在 Postman 中增加“张三”的信息，输入“张三”的 firstName 和 lastName，再在 Postman 的 URL 处输入 http://localhost:8080/people，并选择 POST 方法，结果如图 5-20 所示。在 Postman 的 URL 处输入 http://localhost:8080/people，

并选择 GET 方法，结果如图 5-21 所示。为了查找 lastName 为“三”的所有 people，先在 Postman 的 URL 处输入“http://localhost:8080/people/search/findByLastName?name=三”，并选择 GET 方法，结果如图 5-22 所示。最终处理结果在 MongoDB 数据库可视化管理工具 Robot 3T 中的显示情况如图 5-23 所示。

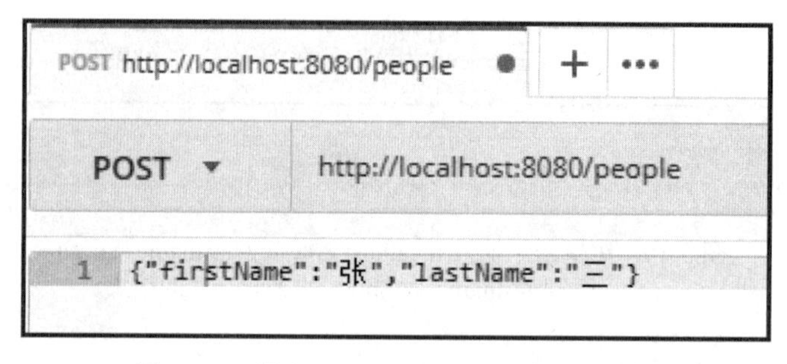

图 5-20　利用 Postman 增加“张三”的信息

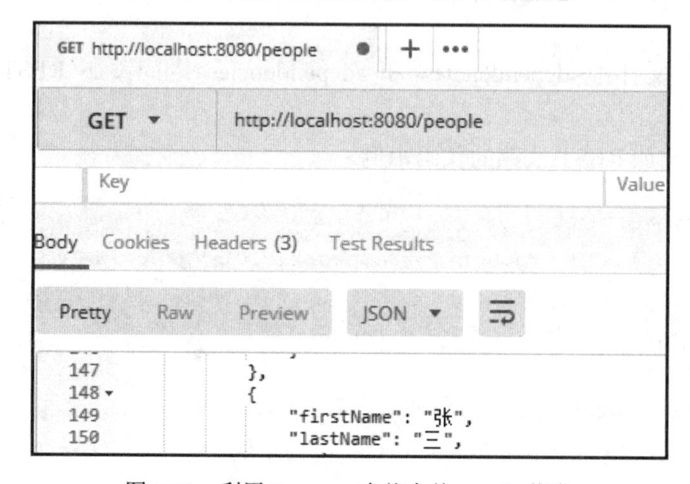

图 5-21　利用 Postman 查找全体 people 信息

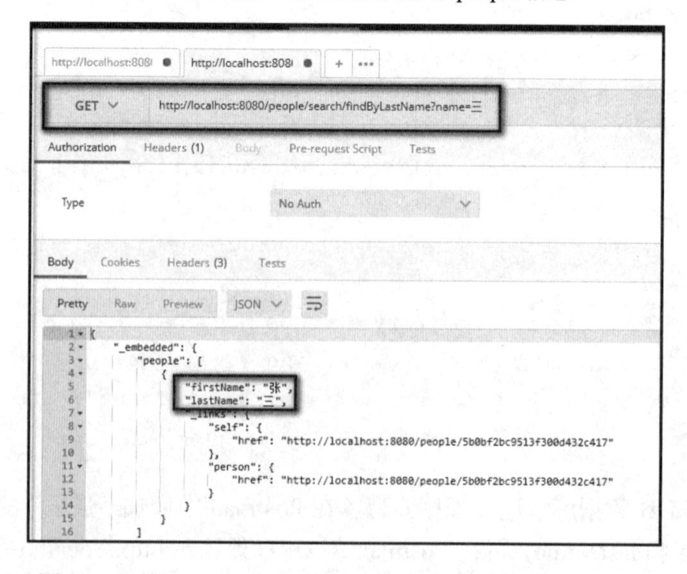

图 5-22　利用 Postman 查找 lastName 为“三”的所有 people

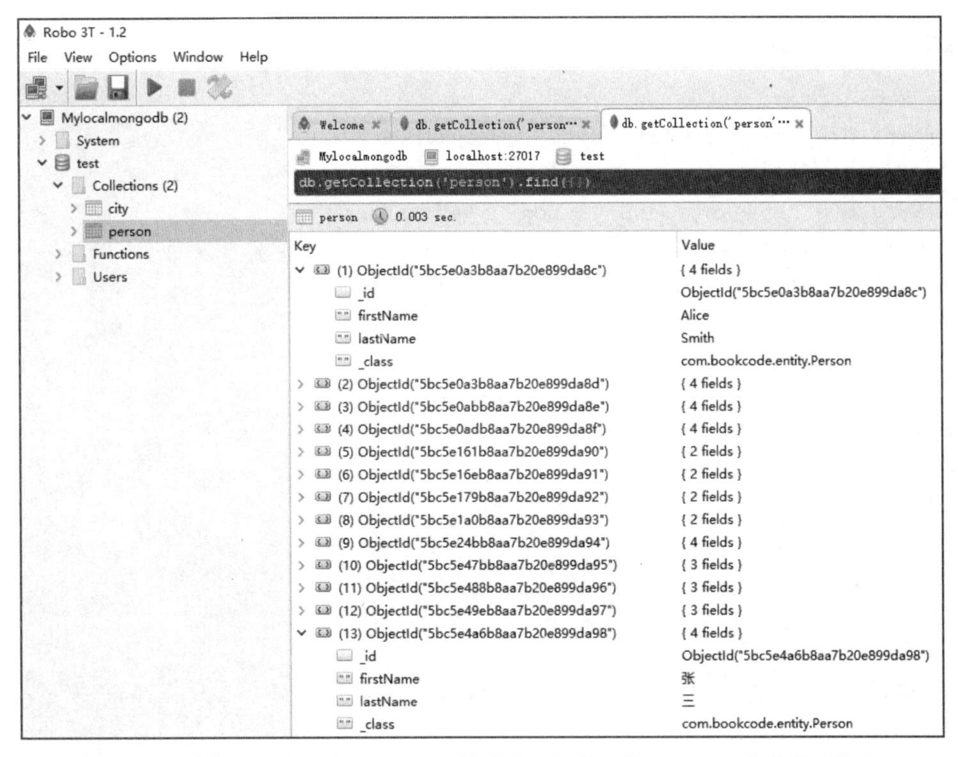

图 5-23　最终处理结果在 MongoDB 数据库可视化工具 Robo 3T 中的显示情况

5.6　访问 Neo4j 数据库

视频讲解

Neo4j 数据库是一个高性能的 NoSQL 图数据库，并且具备完全事务性。Neo4j 数据库将结构化数据存储在一张图上，图中每个结点的属性表示数据的内容，每一条有向边表示数据的关系。Neo4j 数据库的安装过程比较简单，本书不做介绍。

5.6.1　添加依赖

在 pom.xml 文件中<dependencies>和</dependencies>之间添加 Neo4j 依赖，代码如例 5-26 所示。

【例 5-26】 添加 Neo4j 依赖的代码示例。

```
<dependency>
     <groupId>org.springframework.boot</groupId>
     <artifactId>spring-boot-starter-data-neo4j</artifactId>
</dependency>
```

5.6.2　创建类 Actor

在包 com.bookcode.entity 中创建类 Actor，代码如例 5-27 所示。

【例 5-27】 创建类 Actor 的代码示例。

```
package com.bookcode.entity;
import org.neo4j.ogm.annotation.GraphId;
import org.neo4j.ogm.annotation.NodeEntity;
import org.neo4j.ogm.annotation.Relationship;
import java.util.Collections;
import java.util.HashSet;
import java.util.Optional;
import java.util.Set;
import java.util.stream.Collectors;
@NodeEntity                                              //结点实体
public class Actor {
    @GraphId
    private Long id;
    private String name;
    public Actor() {
    }
    public Actor(String name) {
        this.name = name;
    }
@Relationship(type = "TEAMMATE", direction = Relationship.UNDIRECTED)
                                                         //结点间关系
    public Set<Actor> teammates;
    public void worksWith(Actor actor) {
        if (teammates == null) {
            teammates = new HashSet<>();
        }
        teammates.add(actor);
    }
    public String getName() {
        return name;
    }
    public void setName(String name) {
        this.name = name;
    }
    public String toString() {
        return this.name + "'s teammates => "
                + Optional.ofNullable(this.teammates).orElse(
                Collections.emptySet()).stream()
                .map(Actor::getName)
                .collect(Collectors.toList());
    }
}
```

5.6.3　创建接口 ActorRepository

在包 com.bookcode.dao 中创建接口 ActorRepository，代码如例 5-28 所示。

【例 5-28】 创建接口 ActorRepository 的代码示例。

```
package com.bookcode.dao;
import com.bookcode.entity.Actor;
import org.springframework.data.repository.CrudRepository;
public interface ActorRepository extends CrudRepository<Actor,Long> {
    Actor findByName(String name);
}
```

5.6.4　修改配置文件 application.properties

在 application.properties 文件中配置 Neo4j 数据源的用户名和密码，端口采用安装时的默认端口（不用配置），其代码如例 5-29 所示。

【例 5-29】 在 application.properties 文件中配置 Neo4j 数据源的代码示例。

```
#Neo4j 数据源基本配置信息
spring.data.neo4j.username=neo4j
spring.data.neo4j.password=sa
```

5.6.5　修改入口类

修改入口类，代码如例 5-30 所示。

【例 5-30】 修改入口类的代码示例。

```
package com.bookcode;
import com.bookcode.dao.ActorRepository;
import com.bookcode.entity.Actor;
import org.slf4j.Logger;
import org.slf4j.LoggerFactory;
import org.springframework.boot.CommandLineRunner;
import org.springframework.boot.SpringApplication;
import org.springframework.boot.autoconfigure.SpringBootApplication;
import org.springframework.context.annotation.Bean;
import org.springframework.data.neo4j.repository.config.EnableNeo4jRepositories;
import org.springframework.transaction.annotation.EnableTransactionManagement;
import java.util.Arrays;
import java.util.List;
@EnableNeo4jRepositories
@SpringBootApplication
public class DemoApplication {
```

```
    private final static Logger log = LoggerFactory.getLogger(DemoApplication.class);
    public static void main(String[] args) {
        SpringApplication.run(DemoApplication.class, args);
    }
    @Bean
    CommandLineRunner demo(ActorRepository repository) {
        return args -> {
            repository.deleteAll();            //辅助工作，删除所有信息
            Actor greg = new Actor("Greg");
            Actor roy = new Actor("Roy");
            Actor craig = new Actor("Craig");
            List<Actor> team = Arrays.asList(greg, roy, craig);
            log.info("Before linking up with Neo4j...");
            team.stream().forEach(Actor -> log.info("\t" + Actor.toString()));
            repository.save(greg);
            repository.save(roy);
            repository.save(craig);
            greg = repository.findByName(greg.getName());
            greg.worksWith(roy);
            greg.worksWith(craig);
            repository.save(greg);
            roy =repository.findByName(roy.getName());
            roy.worksWith(craig);
            repository.save(roy);
            log.info("Lookup each Actor by name...");
            team.stream().forEach(Actor -> log.info(
                    "\t" + repository.findByName(Actor.getName()).toString()));
        };
    }
}
```

5.6.6 运行程序

先在 Windows 命令处理程序 CMD 中输入命令，启动 Neo4j 数据库，代码如例 5-31 所示。代码中的路径和 Neo4j 数据库安装时的路径有关。

【例 5-31】 启动 Neo4j 数据库命令的代码示例。

```
cd D:\neo4j-community-3.4.0\bin
d:
neo4j console
```

运行程序，在浏览器中输入 http://localhost:7474，可以在 Neo4j 数据库的控制台中看到对 Neo4j 数据库的处理情况，结果如图 5-24 所示。

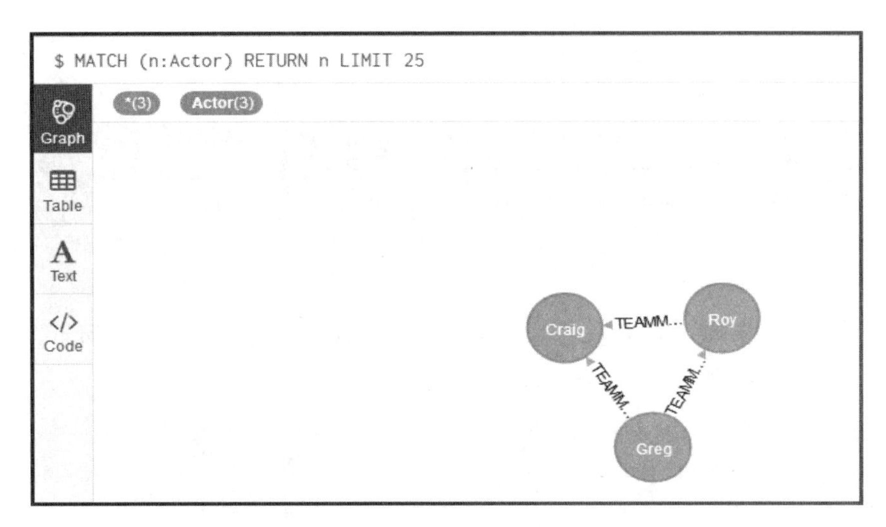

图 5-24　在浏览器中输入 http://localhost:7474 后在 Neo4j 控制台中看到的对 Neo4j 数据库的处理结果

5.6.7　利用 REST 方法访问 Neo4j

在 pom.xml 文件中<dependencies>和</dependencies>之间添加 REST 依赖，代码如例 5-32 所示。

【例 5-32】 添加 REST 依赖的代码示例。

```
<dependency>
         <groupId>org.springframework.boot</groupId>
         <artifactId>spring-boot-starter-data-rest</artifactId>
</dependency>
```

在包 com.bookcode.entity 中创建类 Person，代码如例 5-33 所示。

【例 5-33】 创建类 Person 的代码示例。

```
package com.bookcode.entity;
import org.neo4j.ogm.annotation.GraphId;
import org.neo4j.ogm.annotation.NodeEntity;
@NodeEntity
public class Person {
   @GraphId
   private Long id;
   private String firstName;
   private String lastName;
   public String getFirstName() {
      return firstName;
   }
   public void setFirstName(String firstName) {
      this.firstName = firstName;
   }
```

```
    public String getLastName() {
        return lastName;
    }
    public void setLastName(String lastName) {
        this.lastName = lastName;
    }
}
```

在包 com.bookcode.dao 中创建接口 PersonRepository，代码如例 5-34 所示。

【例 5-34】 创建接口 PersonRepository 的代码示例。

```
package com.bookcode.dao;
import com.bookcode.entity.Person;
import org.springframework.data.repository.PagingAndSortingRepository;
import org.springframework.data.repository.query.Param;
import org.springframework.data.rest.core.annotation.RepositoryRestResource;
import java.util.List;
@RepositoryRestResource(collectionResourceRel = "people", path = "people")
//路径
public  interface  PersonRepository  extends  PagingAndSortingRepository
<Person, Long> {
    List<Person> findByLastName(@Param("name") String name);
}
```

在配置文件中增加 Neo4j 的配置信息，代码如例 5-29 所示。

修改入口类，代码如例 5-35 所示。

【例 5-35】 修改入口类的代码示例。

```
package com.bookcode;
import org.springframework.boot.SpringApplication;
import org.springframework.boot.autoconfigure.SpringBootApplication;
import org.springframework.data.neo4j.repository.config.EnableNeo4jRepositories;
import org.springframework.transaction.annotation.EnableTransactionManagement;
@EnableTransactionManagement
@EnableNeo4jRepositories
@SpringBootApplication
public class SpringbootDatarestApplication {
    public static void main(String[] args) {
        SpringApplication.run(SpringbootDatarestApplication.class, args);
    }
}
```

启动 Neo4j 数据库后运行程序。为了查询所有 people 信息，需要在 Postman 的 URL 处输入 http://localhost:8080/people，并选择 GET 方法，结果如图 5-25 所示。增加“孙权”的记录时，先输入“孙权”的 firstName 和 lastName，再在 Postman 的 URL 处输入 http://localhost:8080/people，并选择 POST 方法，结果如图 5-26 所示。为了查找 lastName

为“三”的所有 people，先在 Postman 的 URL 处输入“http://localhost:8080/people/search/findByLastName?name=三”，并选择 GET 方法，结果如图 5-27 所示。由于此时 Neo4j 数据库中无符合条件的记录，因此返回的结果为空。处理结果在 Neo4j 数据库控制台中的显示情况如图 5-28 所示。

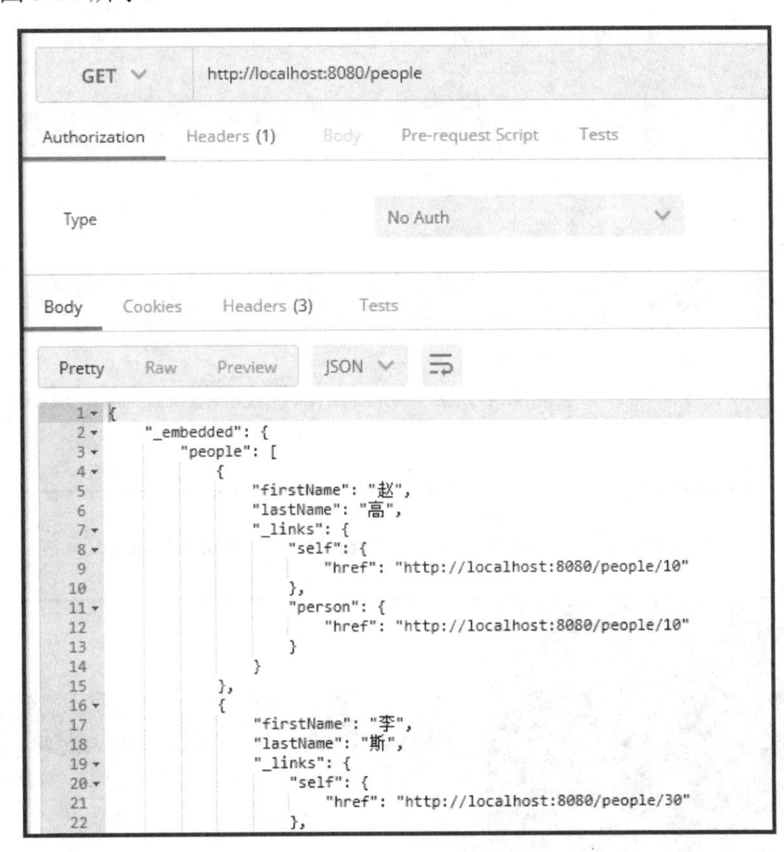

图 5-25　利用 Postman 查询所有 people 信息

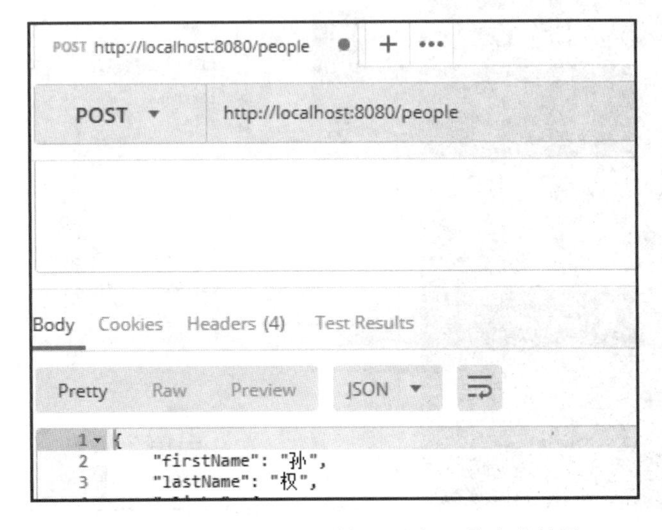

图 5-26　利用 Postman 增加“孙权”信息的结果

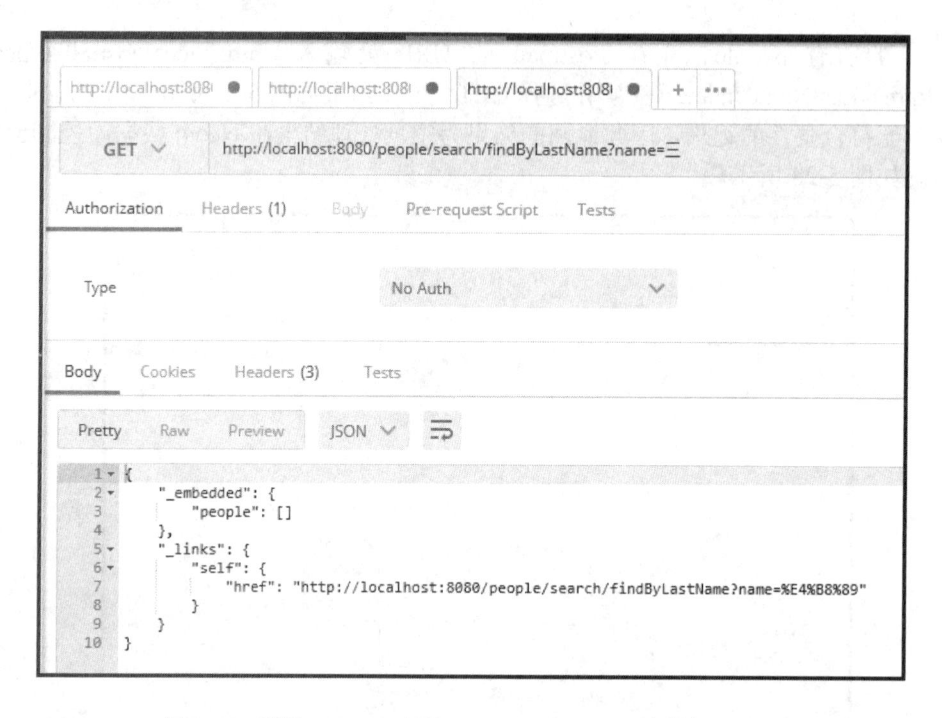

图 5-27 利用 Postman 查找 lastName 为“三”的所有 people

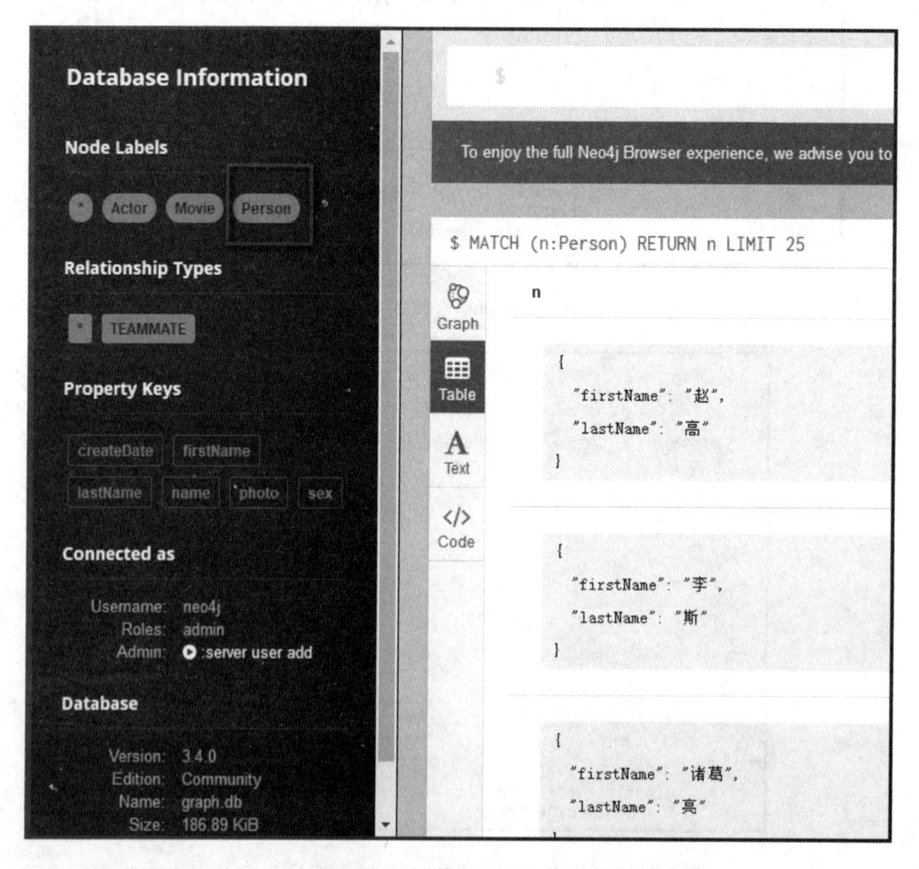

图 5-28 处理结果在 Neo4j 数据库控制台中的显示情况

5.7　访问数据库完整示例

视频讲解

5.7.1　添加依赖

在 pom.xml 文件中<dependencies>和</dependencies>之间添加 Data JPA、Web 等依赖，代码如例 5-36 所示。

【例 5-36】 添加 Data JPA、Web 等依赖的代码示例。

```
<dependency>
        <groupId>org.springframework.boot</groupId>
        <artifactId>spring-boot-starter-data-jpa</artifactId>
</dependency>
<dependency>
        <groupId>org.springframework.boot</groupId>
        <artifactId>spring-boot-starter-thymeleaf</artifactId>
</dependency>
<dependency>
        <groupId>org.springframework.boot</groupId>
        <artifactId>spring-boot-starter-web</artifactId>
</dependency>
<dependency>
        <groupId>mysql</groupId>
        <artifactId>mysql-connector-java</artifactId>
</dependency>
```

5.7.2　创建类 Book

在包 com.bookcode.entity 中创建类 Book，代码如例 5-37 所示。

【例 5-37】 创建类 Book 的代码示例。

```
package com.bookcode.entity;
import javax.persistence.*;
@Entity
@Table(name="t_book")
public class Book {
   @Id
   @GeneratedValue(strategy = GenerationType.IDENTITY)
   private Integer id;
   @Column(length=100)
   private String name;
   @Column(length=50)
   private String author;
   public Integer getId() {
```

```
        return id;
    }
    public void setId(Integer id) {
        this.id = id;
    }
    public String getName() {
        return name;
    }
    public void setName(String name) {
        this.name = name;
    }
    public String getAuthor() {
        return author;
    }
    public void setAuthor(String author) {
        this.author = author;
    }
}
```

5.7.3 创建接口 BookDao

在包 com.bookcode.dao 中创建接口 BookDao，代码如例 5-38 所示。

【例 5-38】 创建接口 BookDao 的代码示例。

```
package com.bookcode.dao;
import com.bookcode.entity.Book;
import org.springframework.data.jpa.repository.JpaRepository;
import org.springframework.data.jpa.repository.JpaSpecificationExecutor;
import org.springframework.data.jpa.repository.Query;
import java.util.List;
public interface BookDao extends JpaRepository<Book, Integer>, JpaSpecification-
Executor<Book> {
    @Query(value = "select * from t_book where t_book.name like  %?1% ",
    nativeQuery = true)
    //或者@Query("select b from Book b where b.name like %?1%")
    public List<Book> findByName(String name);
    //nativeQuery 默认是 HQL 查询，true 表示使用本地查询，就是原生的 SQL 方式
    @Query(value = "select * from t_book  ORDER BY RAND( ) limit ?1 ",
    nativeQuery = true)
    public List<Book> randomList(Integer id);
}
```

5.7.4 修改配置文件 application.properties

修改配置文件 application.properties，代码如例 5-39 所示。

【例 5-39】　修改配置文件 application.properties 的代码示例。

```
spring.datasource.dbcp2.driver-class-name=com.mysql.jdbc.Driver
spring.datasource.url=jdbc:mysql://localhost:3306/mytest
spring.datasource.username=root
spring.datasource.password=sa
spring.jpa.hibernate.ddl-auto=update
spring.jpa.show-sql=true
```

也可以将 application.properties 文件改写成 application.yml，代码如例 5-40 所示。

【例 5-40】　改写配置文件 application.yml 的代码示例。

```
spring:
  datasource:
    driver-class-name: com.mysql.jdbc.Driver
    url: jdbc:mysql://localhost:3306/mytest
    username: root
    password: sa
  jpa:
    hibernate:
      ddl-auto: update
    show-sql: true
```

5.7.5　创建类 BookController

在包 com.bookcode.controller 中创建类 BookController，代码如例 5-41 所示。

【例 5-41】　创建类 BookController 的代码示例。

```
package com.bookcode.controller;
import com.bookcode.dao.BookDao;
import com.bookcode.entity.Book;
import org.springframework.data.jpa.domain.Specification;
import org.springframework.stereotype.Controller;
import org.springframework.web.bind.annotation.*;
import org.springframework.web.servlet.ModelAndView;
import javax.annotation.Resource;
import javax.persistence.criteria.CriteriaBuilder;
import javax.persistence.criteria.CriteriaQuery;
import javax.persistence.criteria.Root;
import java.util.List;
import javax.persistence.criteria.Predicate;
//图书控制器
@Controller
@RequestMapping("/book")
public class BookController {
    @Resource
```

```
private BookDao bookDao;
//查询所有图书
@RequestMapping("/list")
public ModelAndView list(){
    ModelAndView mav=new ModelAndView();
    mav.addObject("booklist", bookDao.findAll());
    mav.setViewName("bookList");
    return mav;
}
//添加图书
@RequestMapping(value="/add",method=RequestMethod.POST)
public String add(Book book){
    bookDao.save(book);
    return "forward:/book/list";
}
//根据 id 查询 book 实体
@RequestMapping("/preUpdate/{id}")
public ModelAndView preUpdate(@PathVariable("id")Integer id){
    ModelAndView mav=new ModelAndView();
    mav.addObject("book", bookDao.getOne(id));
    mav.setViewName("bookUpdate");
    return mav;
}
//修改图书
@PostMapping(value="/update")
public String update(Book book){
    bookDao.save(book);
    return "forward:/book/list";
}
//删除图书
@GetMapping("/delete")
public String delete(Integer id){
    bookDao.deleteById(id);
    return "forward:/book/list";
}
//根据条件动态查询
@RequestMapping("/list2")
public ModelAndView list2(Book book){
    ModelAndView mav = new ModelAndView();
    List<Book> bookList = bookDao.findAll(new Specification<Book>(){
        @Override
     public  Predicate  toPredicate(Root<Book>  root,  CriteriaQuery<?>
     query, CriteriaBuilder cb)
{
        Predicate predicate = cb.conjunction();
```

```
            if(book!=null){
                if(book.getName()!=null && !"".equals(book.getName())){
            predicate.getExpressions().add(cb.like(root.get("name"), "%"+
            book.getName()+"%"));
                }
                if(book.getAuthor()!=null && !"".equals(book.getAuthor())){
           predicate.getExpressions().add(cb.like(root.get("author"), "%"+
           book.getAuthor()+"%"));
                }
            }
            return predicate;
        }
    });
    mav.addObject("book", book);
    mav.addObject("booklist", bookList);
    mav.setViewName("booklist");
    return mav;
  }
  //查询
  @ResponseBody
  @RequestMapping("/query")
  public List<Book> findByName(String name){
      return  bookDao.findByName("思想");
  }
  //随机显示
  @ResponseBody
  @RequestMapping("/randomlist")
  public List<Book> randomList(String name){
      return  bookDao.randomList(2);
  }
}
```

5.7.6　创建文件 bookAdd.html

在 resources/static 目录下创建文件 bookAdd.html，代码如例 5-42 所示。

【例 5-42】 创建文件 bookAdd.html 的代码示例。

```
<!DOCTYPE html>
<html>
<head>
<meta charset="UTF-8">
<title>Insert title here</title>
</head>
<body>
<form action="/book/add" method="post">
 图书名称：<input type="text" name="name"/><br/>
```

```
  图书作者：<input type="text" name="author"/><br/>
  <input type="submit" value="提交"/>
</form>
</body>
</html>
```

5.7.7 创建文件 bookList.html

在 resources/templates 目录下创建文件 bookList.html，代码如例 5-43 所示。

【例 5-43】 创建文件 bookList.html 的代码示例。

```
<!DOCTYPE html>
<html xmlns="http://www.w3.org/1999/xhtml" xmlns:th="http://www.thymeleaf.org" >
<head>
<meta charset="UTF-8">
<title>图书管理</title>
</head>
<body>
<a href="/bookAdd.html">添加</a><br/>
<table>
<tr>
      <th>编号</th>
      <th>图书名称</th>
      <th>图书作者</th>
      <th>操作</th>
</tr>
<p  th:each ="book:${booklist}" >
      <tr>
          <td th:text ="${book.id}"></td>
          <td th:text ="${book.name}"></td>
          <td th:text ="${book.author}"></td>
          <td>
              <a th:href="'/book/preUpdate/'+${book.id}">修改</a>
              <a th:href="'/book/delete?id='+${book.id}">删除</a>
          </td>
      </tr>
</p>
</table>
</body>
</html>
```

5.7.8 创建文件 bookUpdate.html

在 resources/templates 目录下创建文件 bookUpdate.html，代码如例 5-44 所示。

【例 5-44】 创建文件 bookUpdate.html 的代码示例。

```
<!DOCTYPE html>
<html xmlns="http://www.w3.org/1999/xhtml" xmlns:th="http://www.thymeleaf.org" >
<head>
<meta charset="UTF-8">
<title>图书修改</title>
</head>
<body>
<form th:action="'/book/update'" th:method="post">
 <input th:type="hidden" th:name="id" th:value="${book.id}"/>
 图书名称: <input th:type="text" th:name="name" th:value="${book.name}"/><br/>
 图书作者: <input th:type="text" th:name="author" th:value="${book.author}"/>
 <br/>
 <input th:type="submit" th:value="提交"/>
</form>
</body>
</html>
```

5.7.9　运行程序

运行程序后在浏览器中输入 localhost:8080/book/list 或者 localhost:8080/book/list2，结果如图 5-29 所示。进行删除操作时（以删除第 6 条记录为例），在浏览器中输入 localhost:8080/book/delete?id=6，结果如图 5-30 所示。单击“添加”链接后，浏览器中显示为 localhost:8080/bookAdd.html，结果如图 5-31 所示。输入图书名称、图书作者后单击“提交”按钮，结果如图 5-32 所示。单击第 12 条图书记录后面的“修改”链接后，结果如图 5-33 所示。输入修改的第 12 条图书记录信息后单击“提交”按钮，结果如图 5-34 所示。在浏览器中输入 localhost:8080/book/query 后输出的是图书名称中含“思想”的图书信息，结果如图 5-35 所示。在浏览器中输入 localhost:8080/book/randomlist 后随机得到的两本图书记录信息，结果如图 5-36 所示。

图 5-29　在浏览器中输入 localhost:8080/book/list（显示所有图书）的结果

图 5-30 在浏览器中输入 localhost:8080/book/delete?id=6（删除第 6 条记录）的结果

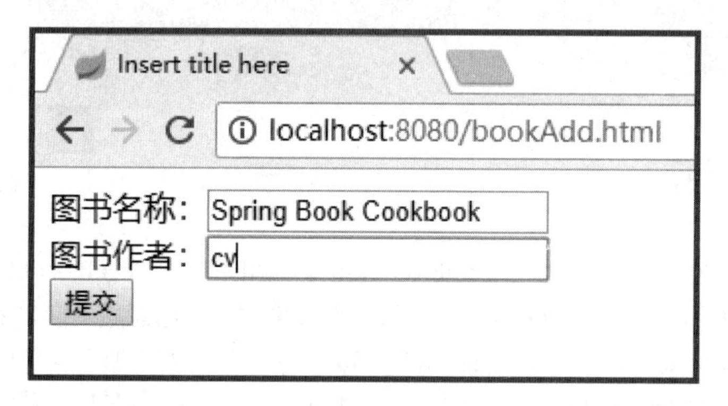

图 5-31 单击“添加”链接后跳转到添加图书信息的界面

图 5-32 输入图书名称、图书作者后单击“提交”按钮（成功添加图书）的结果

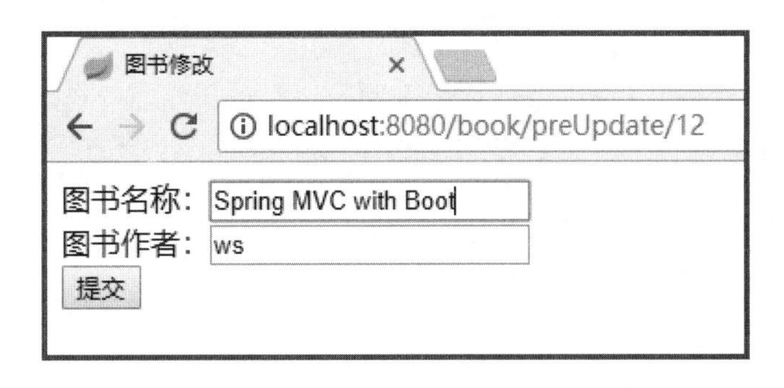

图 5-33　单击第 12 条图书记录后面的“修改”链接后（修改第 12 条记录）的结果

图 5-34　输入修改的第 12 条图书记录信息后单击“提交”按钮
（成功修改第 12 条记录）的结果

图 5-35　在浏览器中输入 localhost:8080/book/query 的输出
（所有名称中含“思想”的图书）

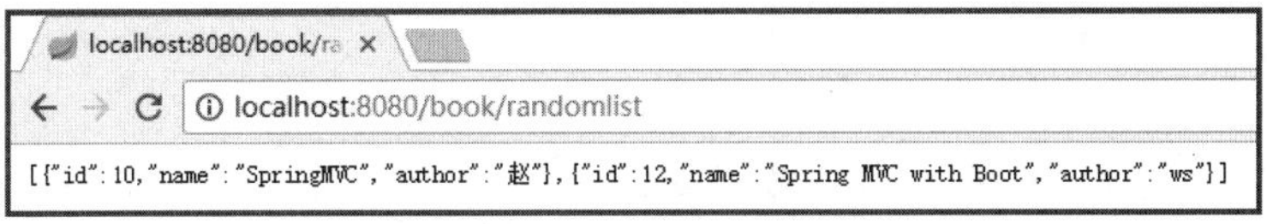

localhost:8080/book/randomlist
[{"id":10,"name":"SpringMVC","author":"赵"},{"id":12,"name":"Spring MVC with Boot","author":"ws"}]

图 5-36　在浏览器中输入 localhost:8080/book/randomlist
后随机得到的两本图书记录信息

习题 5

简答题

1．简述 NoSQL 数据库的特点和适用环境。

2．简述对 JPA 和 Spring Data JPA 的理解。

实验题

1．实现用 JDBC 方式访问 H2 数据库。

2．实现用 Spring Data JPA 访问 H2 数据库。

3．实现用 RESTful 和 Spring Data JPA 访问 H2 数据库。

4．安装工具 Postman，并结合实例使用它。

5．安装数据库 MySQL 和可视化管理工具 Navicat for MySQL，并结合实例使用它。

6．实现使用 Spring Data JPA 方式访问 MySQL 数据库。

7．安装数据库 MongoDB 和可视化管理工具 Robo 3T，并结合实例使用它。

8．实现对 MongoDB 数据库的访问。

9．安装数据库 Neo4j。

10．实现对 Neo4j 数据库的访问。

第 6 章

Spring Boot 的 Web 服务开发

Web 服务（Web Service）是一个平台独立、低耦合、自包含的 Web 应用程序，可使用开放的 XML（可扩展标记语言）标准来描述、发布、发现、协调和配置这些应用程序。Web 服务能使运行在不同机器上的应用无须借助附加的软件或硬件，就可以相互交换数据或集成。只要是依据 Web 服务规范实现的 Web 服务，彼此之间就可以透明地相互交换数据，而不用关注实现的语言、平台或内部协议。Web 服务很容易部署，因为它们基于一些常规产业标准以及已有技术（如 XML、HTTP 等）。Web 服务减少了应用接口的花费。

本章主要介绍如何实现 RESTful 风格 Web 服务，依次介绍如何基于 Jersey 实现 RESTful 风格 Web 服务、如何使用 RESTful 风格 Web 服务、如何使用带 AngularJS 的 RESTful 风格 Web 服务、如何基于 Actuator 实现 RESTful 风格 Web 服务、如何实现跨域资源共享的 RESTful 风格 Web 服务、如何实现超媒体驱动的 RESTful 风格 Web 服务、如何整合 CFX 实现 Web 服务开发。

6.1　基于 Jersey 实现 RESTful 风格 Web 服务

视频讲解

Jersey 框架是开源的 RESTful 框架，实现了 JAX-RS（JSR 311 和 JSR 339）规范。它扩展了 JAX-RS 参考实现， 提供了更多的特性和工具，可以进一步地简化 RESTful 服务和客户端开发。尽管相对年轻，它已经是一个产品级的 RESTful 服务和客户端框架。与 Struts 类似，它同样可以和 Hibernate、Spring 框架整合。

6.1.1　添加依赖

在 pom.xml 文件中<dependencies>和</dependencies>之间添加 Jersey、Fastjson 依赖，代码如例 6-1 所示。

【例 6-1】 添加 Jersey、Fastjson 依赖的代码示例。

```
<dependency>
        <groupId>org.springframework.boot</groupId>
        <artifactId>spring-boot-starter-jersey</artifactId>
</dependency>
<dependency>
        <groupId>com.alibaba</groupId>
        <artifactId>fastjson</artifactId>
        <version>1.2.50</version>
</dependency>
```

6.1.2 创建类 Constant

在 com.bookcode.constant 包中创建类 Constant，代码如例 6-2 所示。

【例 6-2】 创建类 Constant 的代码示例。

```
package com.bookcode.constant;
import java.util.HashMap;
import java.util.Map;
public class Constant {
    public static final String name = "蔡琴";
    public static final Map<String,Object> map;
    static{
        map = new HashMap<String, Object>();
        map.put("歌手", name);
        map.put("歌曲", "被遗忘的时光");
    }
}
```

6.1.3 创建类 JerseyController

在包 com.bookcode.controller 中创建类 JerseyController，代码如例 6-3 所示。

【例 6-3】 创建类 JerseyController 的代码示例。

```
package com.bookcode.controller;
import com.bookcode.constant.Constant;
import org.springframework.web.bind.annotation.RestController;
import javax.ws.rs.*;
import javax.ws.rs.core.MediaType;
import java.util.Map;
import com.alibaba.fastjson.JSON;
//所有注册的端点都应该被@Components 和 HTTP 资源 annotations（如@GET）注解
```

```
@RestController
@Path("/jersey")
public class JerseyController {
    @GET
    @Path("/get")
    @Consumes({MediaType.APPLICATION_XML, MediaType.APPLICATION_JSON})
    @Produces(MediaType.APPLICATION_JSON)
    public Map<String, Object> getMessage() {
        return Constant.map;
    }
//POST 形式在浏览器地址栏中输入请求路径不一定能访问到。推荐用 Fiddler 工具或者 Firefox 浏览
//器插件（poster 或 HttpRequester）
    @POST
    @Path("/post")
    @Consumes({MediaType.APPLICATION_XML, MediaType.APPLICATION_JSON})
    @Produces(MediaType.APPLICATION_JSON)
    public Map<String, Object> postMessage(Map<String,Object> param) {
        System.out.println(JSON.toJSONString(param));
        return Constant.map;
    }
}
```

6.1.4　创建类 JerseyConfig

在包 com.bookcode.config 中创建类 JerseyConfig，代码如例 6-4 所示。

【例 6-4】 创建类 JerseyConfig 的代码示例。

```
package com.bookcode.config;
import com.bookcode.controller.JerseyController;
import org.glassfish.jersey.server.ResourceConfig;
import org.springframework.context.annotation.Configuration;
import javax.ws.rs.ApplicationPath;
/* 想要开始使用 Jersey 2.x 只需要加入依赖，然后需要一个 ResourceConfig 类型的@Bean,
用于注册所有的端点（endpoints,demo 为 JerseyController)。
*/
@Configuration
//Jersey Servlet 将被注册并默认映射到/*。@ApplicationPath 添加到 ResourceConfig
//改变该映射
@ApplicationPath("/rest")
public class JerseyConfig extends ResourceConfig {
    public JerseyConfig() {
        register(JerseyController.class);
    }
}
```

6.1.5 修改入口类

修改入口类，代码如例 6-5 所示。

【例 6-5】 修改入口类的代码示例。

```
package com.bookcode;
import com.bookcode.config.JerseyConfig;
import org.glassfish.jersey.servlet.ServletContainer;
import org.glassfish.jersey.servlet.ServletProperties;
import org.springframework.boot.SpringApplication;
import org.springframework.boot.autoconfigure.SpringBootApplication;
import org.springframework.boot.web.servlet.ServletRegistrationBean;
@SpringBootApplication
public class DemoApplication {
    public ServletRegistrationBean jerseyServlet() {
        ServletRegistrationBean registration = new ServletRegistrationBean(
                new ServletContainer(), "/rest/*");
                                        // rest 资源可以在/rest/*路径下访问
        registration.addInitParameter(ServletProperties.JAXRS_APPLICATION_CLASS,
                JerseyConfig.class.getName());
        return registration;
    }
    public static void main(String[] args) {
        SpringApplication.run(DemoApplication.class, args);
    }
}
```

6.1.6 运行程序

运行程序后，在浏览器中输入 http://localhost:8080/rest/jersey/get，结果是 JSON 格式的字符串，如图 6-1 所示。在 Postman 中选择 POST 方法，在 URL 处输入 http://localhost:8080/rest/jersey/post，并输入 JSON 格式的字符串“{"歌手":"蔡琴","歌曲":"张三的歌"}”，如图 6-2 所示。在 Postman 中完成操作后，在控制台的输出结果如图 6-3 所示。

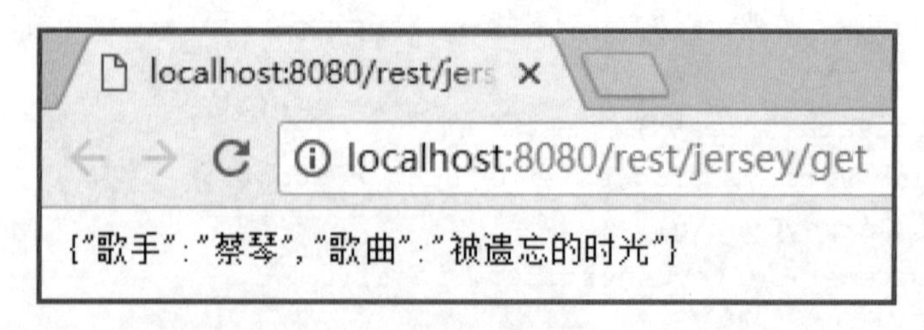

图 6-1 在浏览器中输入 http://localhost:8080/rest/jersey/get 的结果

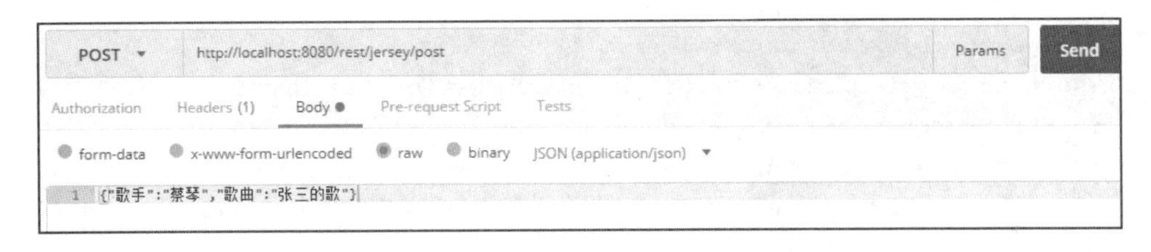

图 6-2　在 Postman 中进行 POST 的相关操作

{"歌手":"蔡琴","歌曲":"张三的歌"}

图 6-3　在 Postman 中完成操作后控制台的输出结果

6.1.7　补充说明

现在前端流行的是静态 HTML+ REST 接口（JSON 格式）。如果是单台服务器，用动态页面还是静态页面可能区别不大，但是如果服务器用到了集群、负载均衡、CDN 等技术，用动态页面还是静态页面差别很大。建议尽量使用 HTML+REST 接口来实现 Web 服务。

REST 接口有两种实现方法：传统实现方法和新的实现方法。传统实现方法的代码如例 6-6 所示。

【例 6-6】 REST 接口传统实现方法的代码示例。

```
package com.bookcode.controller;
import org.springframework.stereotype.Controller;
import org.springframework.web.bind.annotation.RequestMapping;
import org.springframework.web.bind.annotation.RequestMethod;
import org.springframework.web.bind.annotation.ResponseBody;
//注册一个 spring 控制层 Bean
@Controller
public class HelloController {
    //配置方法的 Post 请求 URL
    @RequestMapping(value= "/url/hello",method=RequestMethod.POST)
    //将这个方法的返回值转换成特定格式，默认是 JSON
    @ResponseBody
    //user 用来接收前端请求的相关传参，如 form 表单中的参数
    public String hello(User user){
        return "hello world";
    }
}
```

新的实现方法代码如例 6-7 所示；新的实现方法利用几个新的注解简化了传统的实现方法。

【例 6-7】 REST 接口新的实现方法的代码示例。

```
package com.bookcode.controller;
import org.springframework.web.bind.annotation.PostMapping;
import org.springframework.web.bind.annotation.RequestBody;
import org.springframework.web.bind.annotation.RestController;
//@Controller+@ResponseBody 组合，相当于在每个方法中都加上@ResponseBody
@RestController
public class HiController {
    //直接指定 Post 请求，同样也有@GetMapping
    @PostMapping("/url/hi")
    //@RequestBody 指请求来的参数是 JSON 格式，以 JSON 格式来转换到 user，用 JSON 传
    //参已经越来越多
    public String hi(@RequestBody User user){
        return "hi";
    }
}
```

6.2 使用 RESTful 风格 Web 服务

视频讲解

可以创建应用程序来使用已有的 RESTful 风格 Web 服务，本节介绍如何使用网上（http://gturnquist-quoters.cfapps.io/api/random）已有的 Web 服务。

6.2.1 网上已有 Web 服务 random 的说明

网上 gturnquist-quoters.cfapps.io/api/random 的已有 Web 服务（由于该服务是由 Spring Boot 官方示例引用的，访问者较多，使用时请自行搜索最新网址），可以随机地获取关于 Spring Boot 的引用，通过在浏览器中请求该 URL，可能的结果如图 6-4 所示（结果是随机的）。

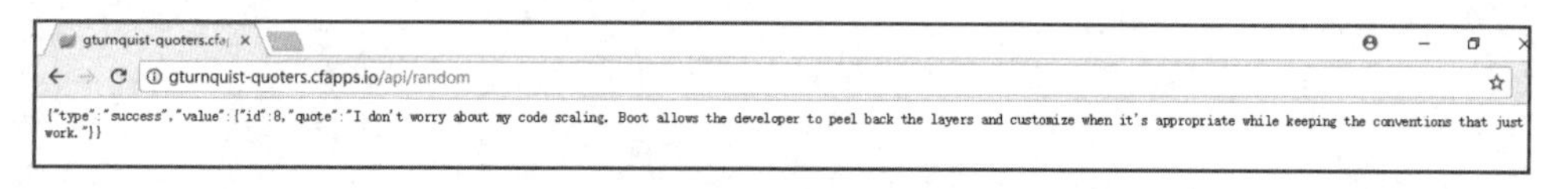

图 6-4 通过浏览器访问 gturnquist-quoters.cfapps.io/api/random 的 random 服务的结果

6.2.2 创建类 Quote

以编程方式可以更方便、有效地使用 RESTful 风格 Web 服务。为了帮助开发者更好地完成该任务，Spring 提供了一个方便的 REST 模板 RestTemplate。RestTemplate 模板使得能以非常少的代码（甚至是一行代码）与大多数 RESTful 风格 Web 服务进行交互，它甚至可以将返回的 JSON 数据绑定到自定义的领域实体类上。于是，使用 RESTful 风格 Web 服务时，需要创建一个用来封装返回 JSON 数据的领域实体类。因此，使用 random 服务时，首先在包 com.bookcode.entity 中创建 Quote 类来封装 random 服务返回的 JSON 数据（如

图 6-4 所示的结果）。类 Quote 的代码如例 6-8 所示。

【例 6-8】 类 Quote 的代码示例。

```
package com.bookcode.entity;
import com.fasterxml.jackson.annotation.JsonIgnoreProperties;
@JsonIgnoreProperties(ignoreUnknown = true)         //忽略类中不存在的字段
public class Quote {
    private String type;
    private Value value;    //需要创建类
    public Quote() {    }
    public String getType() {
        return type;
    }
    public void setType(String type) {
        this.type = type;
    }
    public Value getValue() {
        return value;
    }
    public void setValue(Value value) {
        this.value = value;
    }
    @Override
    public String toString() {
        return "Quote{" +"type='" + type + '\'' +", value=" + value +'}';
    }
}
```

6.2.3　创建类 Value

为了直接将数据绑定到自定义类型，需要指定与 API 返回 JSON 数据中的键完全相同的变量名。如果 JSON 数据中的变量名和密钥不匹配，则需要使用注解@JsonProperties 来忽略类中不存在的字段。为了和返回结果 Quote 类内的属性值对应，需要一个额外的 Value 类。于是，在包 com.bookcode.entity 中创建辅助类 Value，代码如例 6-9 所示。

【例 6-9】 创建辅助类 Value 的代码示例。

```
package com.bookcode.entity;
import com.fasterxml.jackson.annotation.JsonIgnoreProperties;
@JsonIgnoreProperties(ignoreUnknown = true)  //忽略类中不存在的字段
public class Value {
    private Long id;
    private String quote;
    public Value() {    }
    public Long getId() {
```

```
        return this.id;
    }
    public String getQuote() {
        return this.quote;
    }
    public void setId(Long id) {
        this.id = id;
    }
    public void setQuote(String quote) {
        this.quote = quote;
    }
    @Override
    public String toString() {
        return "Value{" +"id=" + id +", quote='" + quote + '\'' +'}';
    }
}
```

6.2.4 修改入口类

修改入口类，代码如例 6-10 所示。

【例 6-10】 修改入口类的代码示例。

```
package com.bookcode
import org.springframework.context.annotation.Bean;
import org.springframework.web.client.RestTemplate;
@SpringBootApplication                         //Spring Boot 应用启动注解
public class SpringbootHelloworldsApplication {
   private static final Logger log = LoggerFactory.getLogger(Springboot
   HelloworldsApplication.class);
    public static void main(String[] args) {
        SpringApplication.run(SpringbootHelloworldsApplication.class,args);
    }
    @Bean                                      //Bean 注解
    public RestTemplate restTemplate(RestTemplateBuilder builder) {
        return builder.build();
    }
    @Bean                                      //Bean 注解
    public CommandLineRunner run(RestTemplate restTemplate) throws Exception {
        return args -> {
Quote quote = restTemplate.getForObject(
  "http://gturnquist-quoters.cfapps.io/api/random", Quote.class);
                                               //用 Quote 封装 random 返回 JSON
    log.info(quote.toString()//将 Quote 信息（返回的 JSON 数据）作为日志记录到控制台
                                                 );
                         };
```

```
        }
}
```

6.2.5　运行程序

访问已经存在的名为 random 的 Web 服务将获得 JSON 数据，模板 RestTemplate 使用位于 classpath 中的 Jackson JSON 处理库将 JSON 数据转换为 Quote 对象。接着，Quote 对象的内容被输出到控制台，结果如图 6-5 所示。

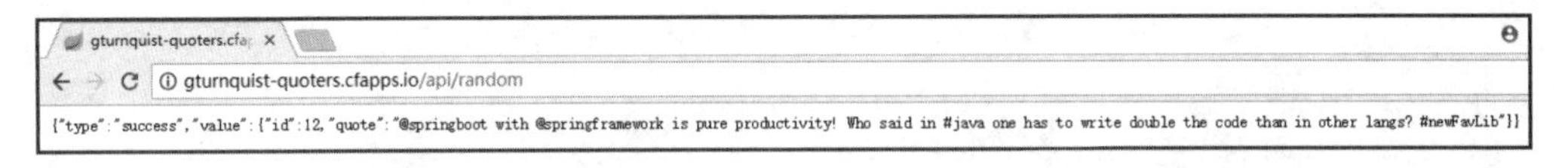

图 6-5　在控制台输出 Quote 对象的内容（由访问名为 random 的 Web 服务获得的 JSON 数据转换而成）

再次，通过浏览器访问（gturnquist-quoters.cfapps.io/api/random）已有 Web 服务 random，可能的结果如图 6-6。对比图 6-4、图 6-5 和图 6-6，可以发现返回的 JSON 数据是随机的。

图 6-6　通过浏览器访问（gturnquist-quoters.cfapps.io/api/random）Web 服务 random 的可能结果

6.3　使用带 AngularJS 的 RESTful 风格 Web 服务

视频讲解

AngularJS 诞生于 2009 年，由 Misko Hevery 等人创建，后被 Google 公司收购。它是一款以 JavaScript 编写的优秀前端 JavaScript 框架，已经被用于 Google 公司的多款产品当中。AngularJS 有着诸多特性，最为核心的包括 MVW（Model-View-Whatever）、模块化、自动化双向数据绑定、语义化标签、依赖注入等。它可通过 <script> </script> 标签添加到 HTML 页面。AngularJS 通过为开发者呈现一个更高层次的抽象来简化应用开发。 AngularJS 主要考虑的是构建 CRUD（Create，增加；Retrieve，查询；Update，更新；Delete，删除）应用。构建一个基于 AngularJS 的 CRUD 应用可能用到的全部内容包括数据绑定、基本模板标识符、表单验证、路由、深度链接、组件重用等。注意，本节使用的服务请读者在官网获取最新网址。

6.3.1　添加依赖和辅助文件

在 pom.xml 文件中<dependencies>和</dependencies>之间添加对 Web 的依赖。下载 AngularJS 并将其添加 resources/static/js 目录下。

6.3.2　创建文件 ajs.html

在 resources/static 目录下创建文件 ajs.html，代码如例 6-11 所示。

【例 6-11】 创建文件 ajs.html 的代码示例。

```
<!doctype html>
<html ng-app="demo">
<head>
    <meta charset="UTF-8"/>
    <title>调用已有的 Web 服务且使用 AngularJS</title>
    <script src="/js/angular.min.js"></script>
    <script>
        angular.module('demo', [])
            .controller('Hello', function($scope, $http) {
                $http.get('http://rest-service.guides.spring.io/greeting').
                then(function(response) {
                    $scope.greeting = response.data;
                });
            });
    </script>
</head>
<body>
<div ng-controller="Hello">
    <h2>调用已有的 Web 服务且使用 AngularJS</h2>
    <p>The ID is {{greeting.id}}</p>
    <p>The content is {{greeting.content}}</p>
</div>
</body>
</html>
```

6.3.3 运行程序

运行程序后，在浏览器中输入 localhost:8080/ajs.html，结果如图 6-7 所示。

图 6-7 在浏览器中输入 localhost:8080/ajs.html 的结果

6.4　基于 Actuator 实现 RESTful 风格 Web 服务

视频讲解

Spring Boot 的 spring-boot-actuator 模块可以对服务做更好的监控和性能查看，可以在应用投入生产时监视和管理应用程序，可以查看服务间的数据处理和调用。当出现异常时，就可以快速定位到出现问题的地方。

6.4.1　添加依赖

在 pom.xml 文件中<dependencies>和</dependencies>之间添加 Actuator、Web 依赖，代码如例 6-12 所示。

【例 6-12】 添加 Actuator、Web 依赖的代码示例。

```
<dependency>
          <groupId>org.springframework.boot</groupId>
          <artifactId>spring-boot-starter-actuator</artifactId>
</dependency>
<dependency>
          <groupId>org.springframework.boot</groupId>
          <artifactId>spring-boot-starter-web</artifactId>
</dependency>
```

6.4.2　创建类 Greeting

在 com.bookcode.entity 包中创建类 Greeting，代码如例 6-13 所示。

【例 6-13】 创建类 Greeting 的代码示例。

```
package com.bookcode.entity;
public class Greeting {
   private final long id;
   private final String content;
   public Greeting(long id, String content) {
      this.id = id;
      this.content = content;
   }
   public long getId() {
      return id;
   }
   public String getContent() {
      return content;
   }
}
```

6.4.3 创建类 GreetingController

创建类 GreetingController，代码如例 6-14 所示。

【例 6-14】 创建类 GreetingController 的代码示例。

```
package com.bookcode.controller;
import com.bookcode.entity.Greeting;
import org.springframework.stereotype.Controller;
import org.springframework.web.bind.annotation.GetMapping;
import org.springframework.web.bind.annotation.RequestParam;
import org.springframework.web.bind.annotation.ResponseBody;
import java.util.concurrent.atomic.AtomicLong;
@Controller
public class GreetingController {
    private static final String template = "Hello, %s!";
    private final AtomicLong counter = new AtomicLong();
    @GetMapping("/hello-world")
    @ResponseBody
    public Greeting sayHello(@RequestParam(name="name", required=false,
    defaultValue="Stranger") String name) {
        return new Greeting(counter.incrementAndGet(), String.format
        (template, name));
    }
}
```

6.4.4 修改配置文件 application.properties

修改配置文件 application.properties，代码如例 6-15 所示。

【例 6-15】 修改配置文件 application.properties 的代码示例。

```
server.port: 9000
management.server.port: 9001
management.server.address: 127.0.0.1
```

6.4.5 运行程序

运行程序后，在浏览器中输入 localhost:9000/hello-world，结果如图 6-8 所示。在浏览器中输入 localhost:9001/actuator/health，结果显示已经启动监控（状态为 UP），如图 6-9 所示。

图 6-8 在浏览器中输入 localhost:9000/hello-world 的结果

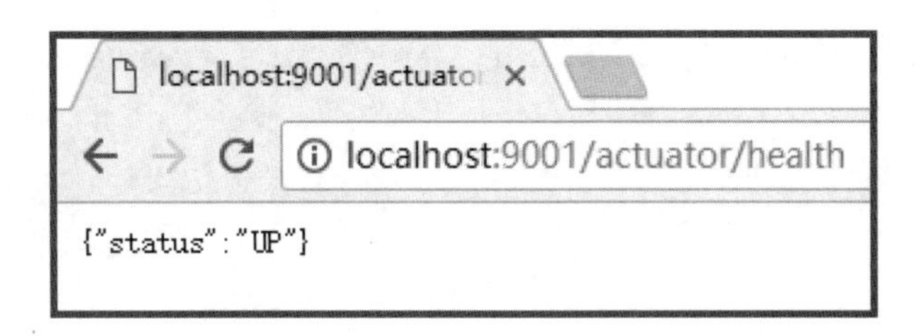

图 6-9　在浏览器中输入 localhost:9001/actuator/health 的结果

6.5　实现跨域资源共享的 RESTful 风格 Web 服务

视频讲解

通过 XMLHttpRequest（XHR）实现 Ajax 通信时一个主要限制就是跨域问题。默认情况下，XHR 对象只能访问与包含它的页面位于同一个域中的资源。跨域资源共享（Cross-Origin Resource Sharing，CORS）解决了 Ajax 跨域的问题。跨域资源共享是一种网络浏览器的技术规范，它为 Web 服务器定义了一种方式，允许网页从不同的域访问其资源。跨域资源共享定义了在必须访问跨域资源时，浏览器与服务器该如何沟通。其背后的基本思想是使用自定义的 HTTP 头部，让浏览器与服务器进行沟通，从而决定请求或响应是应该成功还是应该失败。实现此功能非常简单，只需由服务器发送一个响应确认即可。

6.5.1　添加依赖

在 pom.xml 文件中<dependencies>和</dependencies>之间添加对 Web 的依赖，具体代码请参考 6.4.1 节例 6-12 中第二对<dependency>和</dependency>之间的内容。

6.5.2　创建类 CORSConfiguration

在包 com.bookcode.config 中创建类 CORSConfiguration，代码如例 6-16 所示。

【例 6-16】 创建类 CORSConfiguration 的代码示例。

```
package com.bookcode.config;
import org.springframework.context.annotation.Bean;
import org.springframework.context.annotation.Configuration;
import org.springframework.web.servlet.config.annotation.CorsRegistry;
import org.springframework.web.servlet.config.annotation.WebMvcConfigurer;
import org.springframework.web.servlet.config.annotation.
WebMvcConfigurerAdapter;
@Configuration
public class CORSConfiguration {
    @Bean
    public WebMvcConfigurer corsConfigurer() {
        return new WebMvcConfigurerAdapter() {
            public void addCorsMappings(CorsRegistry registry) {
```

```
                registry.addMapping("/api/**").allowedOrigins("http://
               127.0.0.1:8080");
            }
        };
    }
}
```

6.5.3 创建类 ApiController

在 com.bookcode.controller 包中创建类 ApiController，代码如例 6-17 所示。

【例 6-17】 创建类 ApiController 的代码示例。

```
package com.bookcode.controller;
import org.springframework.web.bind.annotation.RequestMapping;
import org.springframework.web.bind.annotation.RequestParam;
import org.springframework.web.bind.annotation.RestController;
import java.util.HashMap;
@RestController
public class ApiController{
       @RequestMapping(value = "/get")
public HashMap<String, Object> get(@RequestParam  (value="name",required=
false,defaultValue="张三")  String name)
       {
       HashMap<String, Object> hashMap = new HashMap<String, Object>();
       hashMap.put("title", "CORS Test");
       hashMap.put("name", name);
       return hashMap;
       }
}
```

6.5.4 创建文件 CORSjs.html

在 resources/static 目录下创建文件 CORSjs.html，代码如例 6-18 所示。

【例 6-18】 创建文件 CORSjs.html 的代码示例。

```
<!DOCTYPE html>
<html>
<head>
   <meta charset="UTF-8"/>
   <title>使用 CORS</title>
   <script>
      function refreshPrice(data){
         var p=document.getElementById('test-jsonp');
         p.innerHTML='当前价格: '+
```

```
                data['0000001'].name+':'+
                data['0000001'].price+';'+
                data['1399001'].name+':'+
                data['1399001'].price;
        }
        function getPrice(){
            var js=document.createElement('script');
            head=document.getElementsByTagName('head')[0];
            //向跨域网站发送请求资源，并在得到资源后调用 refreshPrice 函数
            js.src='http://api.money.126.net/data/feed/0000001,1399001?
            callback=refreshPrice';
            head.append(js);
        }
    </script>
</head>
<body>
<div>
    <p>CORS 应用——以 163 的股票查询 URL 为例,URL 为 http://api.money.126.net</p>
    <button type="button" onclick="getPrice()">刷新</button>
    <p id="test-jsonp"></p>
</div>
</body>
</html>
```

6.5.5 运行程序

运行程序后，在浏览器中输入 localhost:8080/get，结果如图 6-10 所示。在浏览器中输入 localhost:8080/get?name=%27 李斯%27，结果如图 6-11 所示。在浏览器中输入 localhost:8080/CORSjs.html 并单击“刷新”按钮，结果如图 6-12 所示。

图 6-10 在浏览器中输入 localhost:8080/get 的结果

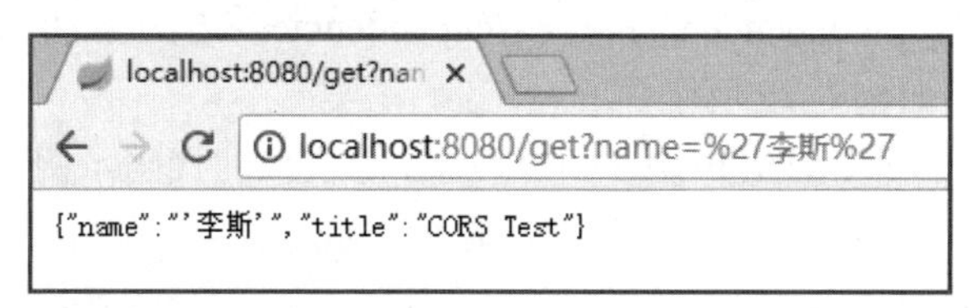

图 6-11 在浏览器中输入 localhost:8080/get?name=%27 李斯%27 的结果

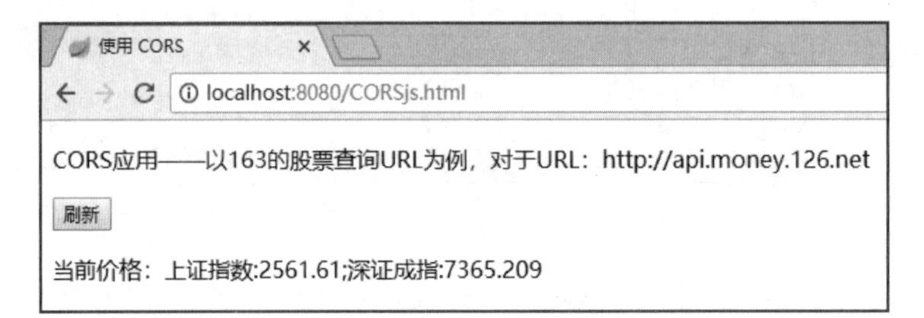

图 6-12　在浏览器中输入 localhost:8080/CORSjs.html 并单击“刷新”按钮后的结果

6.6　实现超媒体驱动的 RESTful 风格 Web 服务

视频讲解

超媒体（Hypermedia）是 REST 的一个重要方面。它允许开发者构建将客户端和服务器解耦并独立演化的服务。为 REST 资源返回的表达（Representations）不仅包含数据，还包含相关资源的链接。因此，表达的设计对于整个服务的设计至关重要。

Richardson 提出的 REST 成熟度模型把 REST 服务按照成熟度划分成四个层次：第一个层次 Web 服务只是使用 HTTP 作为传输方式，实际上只是远程方法调用（RPC）的一种具体形式，SOAP 和 XML-RPC 都属于此类；第二个层次 Web 服务引入了资源的概念，每个资源有对应的标识符和表达；第三个层次 Web 服务使用不同的 HTTP 方法来进行不同操作，并且使用 HTTP 状态码来表示不同结果，如用 HTTP GET 方法来获取资源、用 HTTP DELETE 方法来删除资源；第四个层次 Web 服务使用 HATEOAS（Hypermedia As The Engine Of Application State），在资源的表达中包含了链接信息，客户端可以根据链接来发现可以执行的动作。

从 Richardson 提出的模型中可以看到，使用 HATEOAS 的 REST 服务成熟度最高，也是推荐的做法。它的重要性在于打破了客户端和服务器之间严格的契约，使得客户端可以更加智能和自适应，而 REST 服务本身的演化和更新也变得更加容易。对于不使用 HATEOAS 的 REST 服务，客户端和服务器的实现之间紧密耦合。客户端需要根据服务器提供的相关文档来了解所暴露的资源和对应的操作。当服务器发生了变化时，如修改了资源 URI，客户端也需要进行相应的修改。而使用 HATEOAS 的 REST 服务中，则客户端可以通过服务器提供的资源表达来智能地发现可以执行的操作。若服务器发生了变化，则客户端不需要修改，因为资源的 URI 和其他信息都是动态发现的。

6.6.1　添加依赖

在 pom.xml 文件中<dependencies>和</dependencies>之间添加 Web、HATEOAS 和 Jackson 等依赖，代码如例 6-19 所示。

【例 6-19】 添加 Web、HATEOAS 和 Jackson 等依赖的代码示例。

```
<dependency>
        <groupId>org.springframework.boot</groupId>
        <artifactId>spring-boot-starter-web</artifactId>
```

```
</dependency>
<dependency>
        <groupId>org.springframework.boot</groupId>
        <artifactId>spring-boot-starter-hateoas</artifactId>
</dependency>
<dependency>
        <groupId>com.fasterxml.jackson.core</groupId>
        <artifactId>jackson-annotations</artifactId>
        <version>2.9.0</version>
</dependency>
```

6.6.2　创建类 Greet

在包 com.bookcode.entity 中创建类 Greet，代码如例 6-20 所示。

【例 6-20】 创建类 Greet 的代码示例。

```
package com.bookcode.entity;
import com.fasterxml.jackson.annotation.JsonCreator;
import com.fasterxml.jackson.annotation.JsonProperty;
import org.springframework.hateoas.ResourceSupport;
public class Greet extends ResourceSupport {
   private final String content;
   @JsonCreator
   public Greet(@JsonProperty("content") String content) {
      this.content = content;
   }
   public String getContent() {
      return content;
   }
}
```

6.6.3　创建类 GreetController

在 com.bookcode.controller 包中创建类 GreetController，代码如例 6-21 所示。

【例 6-21】 创建类 GreetController 的代码示例。

```
package com.bookcode.controller;
import static org.springframework.hateoas.mvc.ControllerLinkBuilder.*;
import com.bookcode.entity.Greet;
import org.springframework.http.HttpEntity;
import org.springframework.http.HttpStatus;
import org.springframework.http.ResponseEntity;
import org.springframework.web.bind.annotation.RestController;
import org.springframework.web.bind.annotation.RequestMapping;
```

```
import org.springframework.web.bind.annotation.RequestParam;
@RestController
public class GreetController {
    private static final String TEMPLATE = "Hello, %s!";
    @RequestMapping("/greet")
    public HttpEntity<Greet> greet(
      @RequestParam(value = "name", required = false, defaultValue = "World")
      String name) {
       Greet greet = new Greet(String.format(TEMPLATE, name));
       greet.add(linkTo(methodOn(GreetController.class).greet(name)).
       withSelfRel());
       return new ResponseEntity<>(greet, HttpStatus.OK);
    }
}
```

6.6.4 运行程序

运行程序后，在浏览器中输入 localhost:8080/greet，结果如图 6-13 所示。在浏览器中输入 localhost:8080/greeting?name=zhangsan，结果如图 6-14 所示。

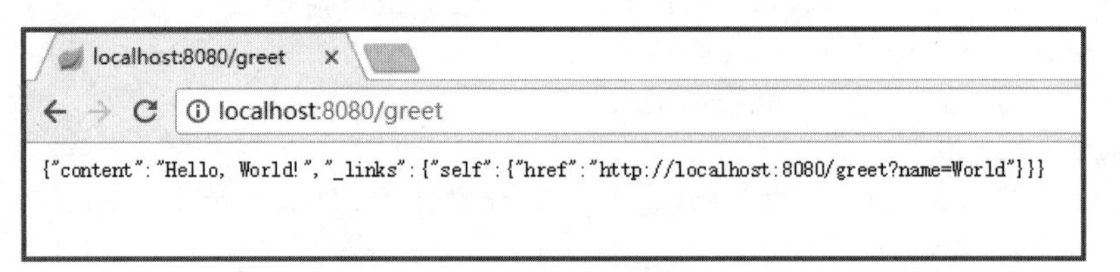

图 6-13　在浏览器中输入 localhost:8080/greet 的结果

图 6-14　在浏览器中输入 localhost:8080/greeting?name=zhangsan 的结果

6.7 整合 CXF 的 Web 服务开发

视频讲解

Apache CXF = Celtix + XFire，开始叫 Apache CeltiXfire，后来更名为 Apache CXF。Apache CXF 是一个开源的 Services 框架。CXF 支持多种 Web Services 标准，包含 SOAP、Basic Profile、WS-Addressing、WS-Policy、WS-ReliableMessaging 和 WS-Security。

SOAP（Simple Object Access Protocol，简单对象访问协议）是交换数据的一种协议规范，是一种简单、轻量、基于 XML 的协议，它被设计成在 Web 上交换结构化和固化的信息。SOAP 和 Web 服务描述语言（Web Services Description Language，WSDL）以及通用描述、发现与集成服务（Universal Description Discovery and Integration，UDDI）是实现 Web 服务的关键。SOAP 用来描述传递信息的格式；WSDL 用来描述如何访问具体的接口；UDDI 用来管理、分发、查询 Web 服务。

6.7.1　修改文件 pom.xml

pom.xml 文件代码如例 6-22 所示。

【例 6-22】 pom.xml 文件的代码示例。

```
<?xml version="1.0" encoding="UTF-8"?>
<project xmlns="http://maven.apache.org/POM/4.0.0" xmlns:xsi="http://www.
w3.org/2001/XMLSchema-instance"
        xsi:schemaLocation="http://maven.apache.org/POM/4.0.0 http:
        //maven.apache.org/xsd/maven-4.0.0.xsd">
    <modelVersion>4.0.0</modelVersion>
    <groupId>com.dbgo</groupId>
    <artifactId>webservicedemo</artifactId>
    <version>0.0.1-SNAPSHOT</version>
    <packaging>jar</packaging>
    <name>webservicedemo</name>
    <description>Demo project for Spring Boot</description>
    <parent>
        <groupId>org.springframework.boot</groupId>
        <artifactId>spring-boot-starter-parent</artifactId>
        <version>1.5.8.RELEASE</version>
        <relativePath/> <!-- lookup parent from repository -->
    </parent>
    <properties>
        <project.build.sourceEncoding>UTF-8</project.build.sourceEncoding>
        <project.reporting.outputEncoding>UTF-8</project.reporting.
        outputEncoding>
        <java.version>1.8</java.version>
    </properties>
    <dependencies>
        <dependency>
            <groupId>org.springframework.boot</groupId>
            <artifactId>spring-boot-starter-web</artifactId>
        </dependency>
        <!--WebService CXF 依赖-->
        <dependency>
            <groupId>org.apache.cxf</groupId>
            <artifactId>cxf-rt-frontend-jaxws</artifactId>
```

```
            <version>3.1.12</version>
        </dependency>
        <dependency>
            <groupId>org.apache.cxf</groupId>
            <artifactId>cxf-rt-transports-http</artifactId>
            <version>3.1.12</version>
        </dependency>
        <dependency>
            <groupId>org.springframework.boot</groupId>
            <artifactId>spring-boot-starter-test</artifactId>
            <scope>test</scope>
        </dependency>
        <dependency>
            <groupId>junit</groupId>
            <artifactId>junit</artifactId>
            <version>4.12</version>
        </dependency>
        <dependency>
            <groupId>commons-io</groupId>
            <artifactId>commons-io</artifactId>
            <version>2.5</version>
        </dependency>
    </dependencies>
    <build>
        <plugins>
            <plugin>
                <groupId>org.springframework.boot</groupId>
                <artifactId>spring-boot-maven-plugin</artifactId>
            </plugin>
        </plugins>
    </build>
</project>
```

6.7.2 创建类 User

在 com.bookcode.entity 中创建类 User，代码如例 6-23 所示。

【例 6-23】 创建类 User 的代码示例。

```
package com.bookcode.entity;
import java.io.Serializable;
import java.util.Date;
public class User implements Serializable {
    private static final long serialVersionUID = -5939599230753662529L;
    private String userId;
    private String username;
```

```
    private String age;
    private Date updateTime;
    public void setUserId(String userId) {
        this.userId=userId;
    }
    public void setUsername(String username) {
        this.username=username;
    }
    public void setAge(String age) {
        this.age=age;
    }
    public void setUpdateTime(Date updateTime) {
        this.updateTime=updateTime;
    }
    public String getUserId() {
        return userId;
    }
    public String getUserName() {
        return username;
    }
    public String getAge() {
        return age;
    }
    public Date getUpdateTime() {
        return updateTime;
    }
}
```

6.7.3 创建接口 UserService

在包 com.bookcode.dao 中创建接口 UserService，代码如例 6-24 所示。

【例 6-24】 创建接口 UserService 的代码示例。

```
package com.bookcode.dao;
import com.bookcode.entity.User;
import javax.jws.WebMethod;
import javax.jws.WebParam;
import javax.jws.WebService;
@WebService
public interface UserService {
    @WebMethod
    String getName(@WebParam(name = "userId") String userId);
    @WebMethod
    User getUser(String userId);
}
```

6.7.4 创建类 UserServiceImpl

在 com.bookcode.dao 包中创建类 UserServiceImpl，代码如例 6-25 所示。

【例 6-25】 创建类 UserServiceImpl 的代码示例。

```
package com.bookcode.dao;
import com.bookcode.entity.User;
import java.util.Date;
import java.util.HashMap;
import java.util.Map;
import javax.jws.WebService;
@WebService(targetNamespace="http://dao.bookcode.com/",endpointInterface=
"com.bookcode.dao.UserService")
public class UserServiceImpl implements UserService{
    private Map<String, User> userMap = new HashMap<String, User>();
    public UserServiceImpl() {
        System.out.println("向实体类插入数据");
        User user = new User();
        user.setUserId("411001");
        user.setUsername("zhansan");
        user.setAge("20");
        user.setUpdateTime(new Date());
        userMap.put(user.getUserId(), user);
        user = new User();
        user.setUserId("411002");
        user.setUsername("lisi");
        user.setAge("30");
        user.setUpdateTime(new Date());
        userMap.put(user.getUserId(), user);
        user = new User();
        user.setUserId("411003");
        user.setUsername("wangwu");
        user.setAge("40");
        user.setUpdateTime(new Date());
        userMap.put(user.getUserId(), user);
    }
    @Override
    public String getName(String userId) {
        return "liyd-" + userId;
    }
    @Override
    public User getUser(String userId) {
        System.out.println("userMap 是:"+userMap);
        return userMap.get(userId);
    }
}
```

6.7.5　创建类 TestConfig

在包 com.bookcode.config 中创建类 TestConfig，代码如例 6-26 所示。

【例 6-26】　创建类 TestConfig 的代码示例。

```
package com.bookcode.config;
import javax.xml.ws.Endpoint;
import com.bookcode.dao.UserService;
import com.bookcode.dao.UserServiceImpl;
import org.apache.cxf.Bus;
import org.apache.cxf.bus.spring.SpringBus;
import org.apache.cxf.jaxws.EndpointImpl;
import org.apache.cxf.transport.servlet.CXFServlet;
import org.springframework.boot.web.servlet.ServletRegistrationBean;
import org.springframework.context.annotation.Bean;
import org.springframework.context.annotation.Configuration;
@Configuration
public class TestConfig {
    @Bean
    public ServletRegistrationBean dispatcherServlet() {
        return new ServletRegistrationBean(new CXFServlet(), "/test/*");
    }
    @Bean(name = Bus.DEFAULT_BUS_ID)
    public SpringBus springBus() {
        return new SpringBus();
    }
    @Bean
    public UserService userService() {
        return new UserServiceImpl();
    }
    @Bean
    public Endpoint endpoint() {
        EndpointImpl endpoint = new EndpointImpl(springBus(), userService());
        endpoint.publish("/user");
        return endpoint;
    }
}
```

6.7.6　运行程序

至此服务器开发完成。运行程序后，在浏览器中输入 localhost:8080/test/user?wsdl，结果如图 6-15 所示。

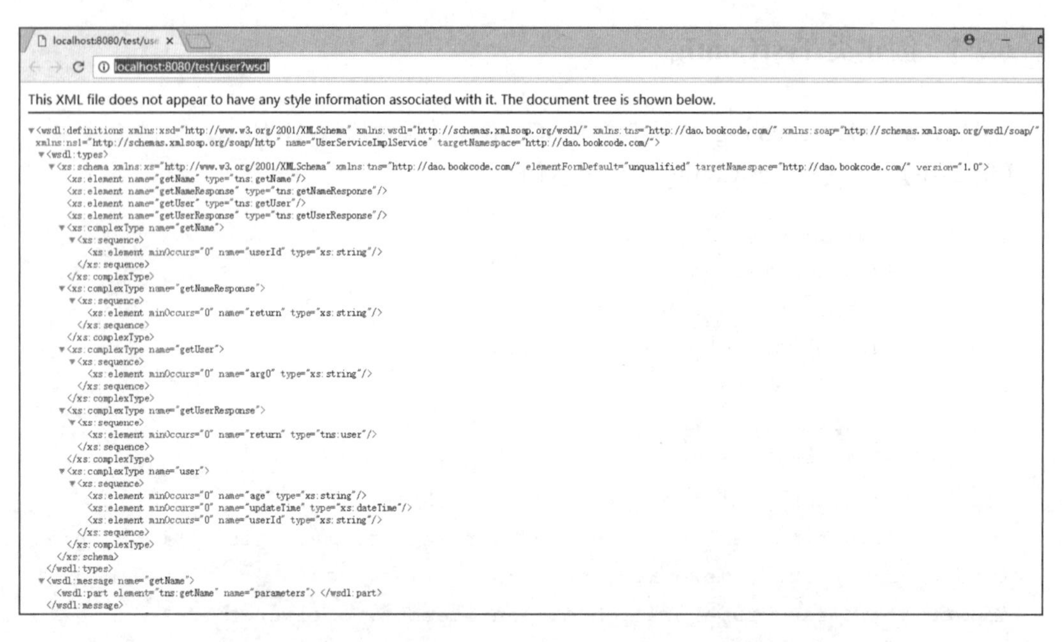

图 6-15 在浏览器中输入 localhost:8080/test/user?wsdl 的结果

6.7.7 创建类 Client 并运行程序

在包 com.bookcode.client 中创建类 Client，代码如例 6-27 所示。

【例 6-27】 创建类 Client 的代码示例。

```
package com.bookcode.client;
import org.apache.cxf.jaxws.endpoint.dynamic.JaxWsDynamicClientFactory;
public class Client {
    public static void main(String args[]) throws Exception{
        JaxWsDynamicClientFactory dcf =JaxWsDynamicClientFactory.newInstance();
        org.apache.cxf.endpoint.Client client =dcf.createClient("http://
        localhost:8080/test/user?wsdl");
        //getUser 为接口中定义的方法名称，张三为传递的参数，返回一个 Object 数组
        Object[] objects=client.invoke("getUser","411001");
        //输出调用结果
        System.out.println("*****"+objects[0].toString()+" 张三，你好！");
    }
}
```

运行 Client 类后，控制台中的输出结果如图 6-16 所示。

```
*****com.bookcode.dao.User@4d9d1b69 张三，你好！
```

图 6-16 运行 Client 后控制台中的输出结果

习题 6

简答题

1．简述对 Web 服务的理解。
2．简述对 Actuator 的理解。
3．简述对跨域资源共享的理解。
4．简述对超媒体驱动和 HATEOAS 的理解。
5．简述对 SOAP 的理解。

实验题

1．基于 Jersey 实现 RESTful 风格 Web 服务。
2．实现对 RESTful 风格 Web 服务的使用。
3．实现对带 AngularJS 的 RESTful 风格 Web 服务的使用。
4．基于 Actuator 实现 RESTful 风格 Web 服务。
5．实现跨域资源共享的 RESTful 风格 Web 服务的使用。
6．实现超媒体驱动的 RESTful 风格 Web 服务。
7．整合 CXF 实现 Web 服务的开发。

第 7 章

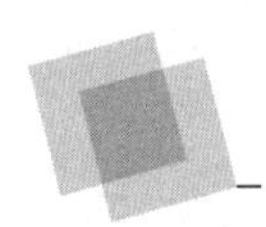

Spring Boot 的数据处理

本章介绍如何实现声明式事务、如何实现数据缓存、如何使用 Druid、如何使用表单验证、如何整合 MyBatis 访问数据库、如何实现对 Spring Batch 和 Quartz 的整合等内容。

7.1 声明式事务

视频讲解

Spring 同时支持编程式事务管理和声明式事务管理。对于编程式事务管理，Spring 推荐使用 TransactionTemplate。声明式事务建立在 AOP 之上，其本质是对方法前后进行拦截，然后在目标方法开始之前创建或者加入一个事务，在执行完目标方法之后根据执行情况提交或者回滚事务。声明式事务就是通过在配置文件中加声明的方式来处理事务的。声明式事务最大的优点就是不需要在业务逻辑代码中掺杂事务管理的代码，只要在配置文件中做相关事务规则的声明（或基于@Transactional 注解的方式）便可以将事务规则应用到业务逻辑中。Spring 使用 AOP 来完成声明式的事务管理，因而声明式事务是以方法为单位的。Spring 的事务属性在于描述事务应用至方法上的策略，在 Spring 中事务属性有传播行为、隔离级别、只读提示、事务超时期间四个参数。和编程式事务相比，声明式事务的最细粒度只能作用到方法级，无法像编程式事务那样可以作用到代码块级。

7.1.1 添加依赖

在 pom.xml 文件中<dependencies>和</dependencies>之间添加依赖，代码如例 7-1 所示。

【例 7-1】 添加依赖的代码示例。

```
<dependency>
        <groupId>org.springframework.boot</groupId>
```

```
        <artifactId>spring-boot-starter-web</artifactId>
</dependency>
<dependency>
        <groupId>mysql</groupId>
        <artifactId>mysql-connector-java</artifactId>
        <scope>runtime</scope>
</dependency>
<dependency>
        <groupId>org.springframework.boot</groupId>
        <artifactId>spring-boot-starter-data-jpa</artifactId>
</dependency>
```

7.1.2　创建类 Account

创建类 Account，代码如例 7-2 所示。

【例 7-2】 创建类 Account 的代码示例。

```
package com.bookcode.entity;
import javax.persistence.Column;
import javax.persistence.Entity;
import javax.persistence.GeneratedValue;
import javax.persistence.Id;
import javax.persistence.Table;
//账户实体
@Entity
@Table(name="t_account")
public class Account {
    @Id
    @GeneratedValue
    private Integer id;
    @Column(length=50)
    private String userName;
    private float balance;
    public Integer getId() {
        return id;
    }
    public void setId(Integer id) {
        this.id = id;
    }
    public String getUserName() {
        return userName;
    }
    public void setUserName(String userName) {
        this.userName = userName;
    }
    public float getBalance() {
```

```
        return balance;
    }
    public void setBalance(float balance) {
        this.balance = balance;
    }
}
```

7.1.3 创建接口 AccountDao

创建接口 AccountDao，代码如例 7-3 所示。

【例 7-3】 创建接口 AccountDao 的代码示例。

```
package com.bookcode.dao;
import com.bookcode.entity.Account;
import org.springframework.data.jpa.repository.JpaRepository;
public interface AccountDao extends JpaRepository<Account, Integer>{
}
```

7.1.4 创建接口 AccountService

创建接口 AccountService，代码如例 7-4 所示。

【例 7-4】 创建接口 AccountService 的代码示例。

```
package com.bookcode.service;
public interface AccountService {
    //从 A 用户转账至 b 用户，金额为 account
    public void transferAccounts(int fromUser, int toUser, float account);
}
```

7.1.5 创建类 AccountController

创建类 AccountController，代码如例 7-5 所示。

【例 7-5】 创建类 AccountController 的代码示例。

```
package com.bookcode.controller;
import javax.annotation.Resource;
import com.bookcode.service.AccountService;
import org.springframework.web.bind.annotation.RequestMapping;
import org.springframework.web.bind.annotation.RestController;
@RestController
@RequestMapping("/account")
public class AccountController {
    @Resource
    private AccountService accoutService;
```

```
    @RequestMapping("/transfer")
    public String transferAccount(){
        try{
            accoutService.transferAccounts(1, 2, 200);
            return "OK";
        }
        catch(Exception e){
            return "NO";
        }
    }
}
```

7.1.6　创建配置文件 application.yml

配置文件 application.yml 的代码如例 7-6 所示。

【例 7-6】 配置文件 application.yml 的代码示例。

```
spring:
  datasource:
    driver-class-name: com.mysql.jdbc.Driver
    url: jdbc:mysql://localhost:3306/mytest
    username: root
    password: sa
  jpa:
    hibernate:
      ddl-auto: update
    show-sql: true
```

7.1.7　创建类 AccountServiceImpl

创建类 AccountServiceImpl，代码如例 7-7 所示。

【例 7-7】 创建类 AccountServiceImpl 的代码示例。

```
package com.bookcode.service.impl;
import javax.annotation.Resource;
import javax.transaction.Transactional;
import com.bookcode.dao.AccountDao;
import com.bookcode.entity.Account;
import com.bookcode.service.AccountService;
import org.springframework.stereotype.Service;
@Service("accountService")
public class AccountServiceImpl implements AccountService {
    @Resource
    private AccountDao accountDao;
    @Transactional
```

```
    public void transferAccounts(int fromUser, int toUser, float account) {
        Account fromAccount=accountDao.getOne(fromUser);
        fromAccount.setBalance(fromAccount.getBalance()-account);
        accountDao.save(fromAccount);
        Account toAccount=accountDao.getOne(toUser);
        toAccount.setBalance(toAccount.getBalance()+account);
        int zero=1/0;
        accountDao.save(toAccount);
    }
}
```

7.1.8 运行程序

运行程序后，在浏览器中输入 localhost:8080/account/transfer，转账操作出现异常，在浏览器中输出 NO，结果如图 7-1 所示。此时的数据库表 t_account 不会发生变化，原始数据在工具 Navicat for MySQL 中的结果如图 7-2 所示。对例 7-7 代码中语句“int zero=1/0;”添加注释后，重新运行程序，在浏览器中输入 localhost:8080/account/transfer，完成正常的转账，浏览器输出 OK，结果如图 7-3 所示。与此同时，数据库表 t_account 中第 1 条记录（张三）、第 2 条记录（李四）的账户 balance 分别减少、增加 200，数据库最新数据在工具 Navicat for MySQL 中的结果如图 7-4 所示。对例 7-7 代码中注解@Transactional 添加注释且去掉对语句“int zero=1/0;”的注释后，重新运行程序，在浏览器中输入 localhost:8080/account/transfer，转账操作出现异常，浏览器输出 NO，结果如图 7-1 所示。与此同时，数据库表 t_account 中第 1 条记录（张三）的账户 balance 减少了 200，但是第 2 条记录（李四）的账户 balance 没有发生变化，转账操作出现异常时数据变动在工具 Navicat for MySQL 中的结果如图 7-5 所示。

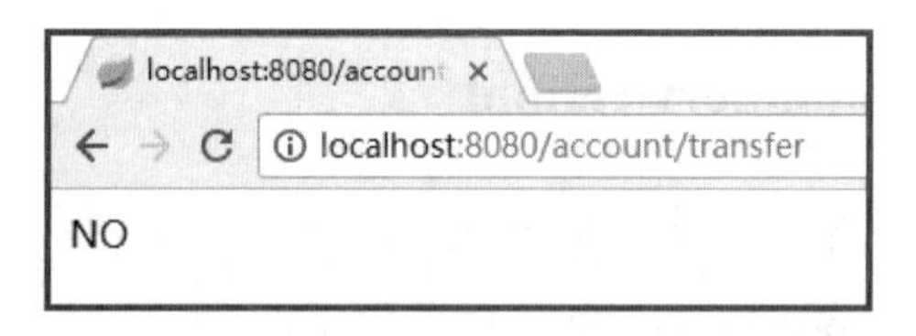

图 7-1 在浏览器中输入 localhost:8080/account/transfer 的结果
（由于执行“1 除 0”而输出 NO）

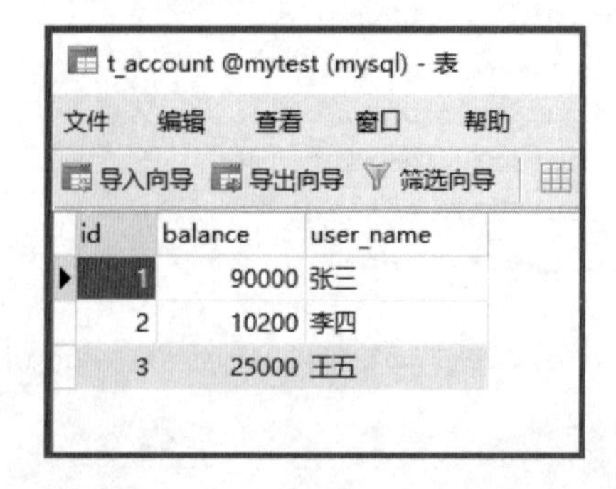

图 7-2 原始数据在工具 Navicat for MySQL 中的结果

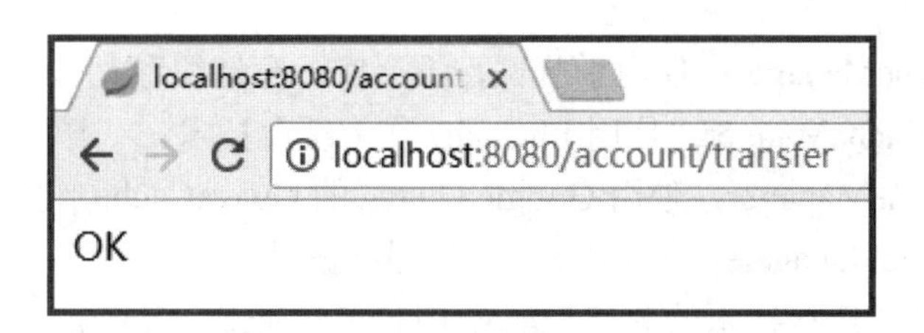

图 7-3　完成正常转账后的输出

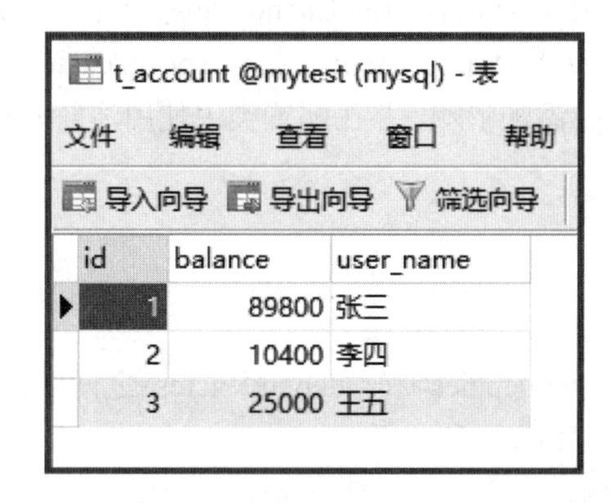

t_account @mytest (mysql) - 表

文件　编辑　查看　窗口　帮助

导入向导　导出向导　筛选向导

id	balance	user_name
1	89800	张三
2	10400	李四
3	25000	王五

图 7-4　完成正常转账后数据库最新数据在工具 Navicat for MySQL 中的结果

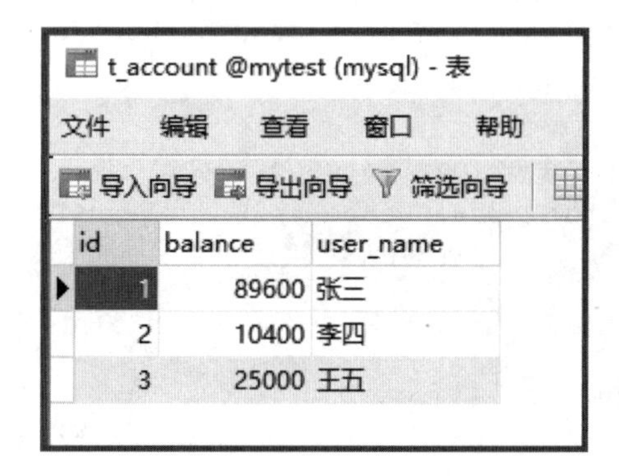

t_account @mytest (mysql) - 表

文件　编辑　查看　窗口　帮助

导入向导　导出向导　筛选向导

id	balance	user_name
1	89600	张三
2	10400	李四
3	25000	王五

图 7-5　转账操作出现异常时数据变动在工具 Navicat for MySQL 中的结果

7.2　数据缓存

视频讲解

在实际开发中，对于要反复读写的数据，好的处理方式是将之在内存中缓存一份。Spring 在数据缓存方面提供了许多处理手段，Spring Boot 又将方式进一步简化。Spring 在 org.springframework.cache 包中定义了 CacheManager 和 Cache 两个接口来统一不同的缓存技术。其中，CacheManager 是 Spring 提供的各种缓存技术抽象接口，Cache 接口包含缓存的各种操作（增加、删除、获得缓存）。针对不同的缓存技术，需要实现不同的 CacheManager，Spring 定义了如下 CacheManager 来实现。

（1）SimpleCacheManager：使用简单的 Collection 来存储缓存，主要用来测试用途。

（2）ConcurrentMapCacheManager：使用 ConcurrentMap 来存储缓存。

（3）NoOpCacheManager：仅用于测试，不会实际存储缓存。

（4）EhCacheCacheManger：使用 EhCache 作为缓存技术。

（5）GuavaCacheManager：使用 Google Guava 的 GuavaCache 作为缓存技术。

（6）RedisCacheManager：使用 Redis 作为缓存技术。

在 Spring 中使用缓存技术的关键是配置 CacheManager，Spring Boot 自动配置了多个 CacheManager 来实现。Spring Boot 的 CacheManager 自动配置放置在 org.springframework.boot.autoconfigure.cache 包中，默认使用的 CacheManager 是 SimpleCacheConfiguration（即 ConcurrentMapCacheManager）。Spring Boot 支持以 spring.cache 为前缀的属性来配置缓存。使用缓存技术只需在项目中添加依赖，并在配置类中使用注解@EnableCaching 开启缓存支持即可。

7.2.1 添加依赖

在 pom.xml 文件中<dependencies>和</dependencies>之间添加依赖，代码如例 7-8 所示。

【例 7-8】 添加依赖的代码示例。

```
<dependency>
        <groupId>org.springframework.boot</groupId>
        <artifactId>spring-boot-starter-web</artifactId>
</dependency>
<dependency>
        <groupId>javax.persistence</groupId>
        <artifactId>persistence-api</artifactId>
        <version>1.0</version>
</dependency>
<dependency>
        <groupId>mysql</groupId>
        <artifactId>mysql-connector-java</artifactId>
</dependency>
<dependency>
        <groupId>org.springframework.boot</groupId>
        <artifactId>spring-boot-starter-data-jpa</artifactId>
</dependency>
```

7.2.2 创建类 DemoInfo

创建类 DemoInfo，代码如例 7-9 所示。

【例 7-9】 创建类 DemoInfo 的代码示例。

```
package com.bookcode.entity;
import javax.persistence.Entity;
import javax.persistence.GeneratedValue;
import javax.persistence.Id;
```

```
@Entity
public class DemoInfo {
    @Id
    @GeneratedValue
    private Long id;          //主键
    private String name;      //名称
    private String pwd;       //密码
    private Integer state;
    public Long getId() {
        return id;
    }
    public void setId(Long id) {
        this.id = id;
    }
    public Integer getState() {
        return state;
    }
    public void setState(Integer state) {
        this.state = state;
    }
    public String getName() {
        return name;
    }
    public void setName(String name) {
        this.name = name;
    }
    public String getPwd() {
        return pwd;
    }
    public void setPwd(String pwd) {
        this.pwd = pwd;
    }
    @Override
    public String toString() {
        return"DemoInfo [id=" + id + ", name=" + name + ", pwd=" + pwd + ",
        state=" + state + "]";
    }
}
```

7.2.3　创建接口 DemoInfoRepository

创建接口 DemoInfoRepository，代码如例 7-10 所示。

【例 7-10】 创建接口 DemoInfoRepository 的代码示例。

```
package com.bookcode.dao;
```

```
import com.bookcode.entity.DemoInfo;
import org.springframework.data.repository.CrudRepository;
public interface DemoInfoRepository extends CrudRepository<DemoInfo,Long> {
}
```

7.2.4 创建接口 DemoInfoService

创建接口 DemoInfoService，代码如例 7-11 所示。

【例 7-11】 创建接口 DemoInfoService 的代码示例。

```
package com.bookcode.service;
import com.bookcode.entity.DemoInfo;
import javassist.NotFoundException;
import java.util.Optional;
public interface DemoInfoService {
    void delete(Long id);
    DemoInfo update(DemoInfo updated) throws NotFoundException;
    Optional<DemoInfo> findById(Long id);
    DemoInfo save(DemoInfo demoInfo);
}
```

7.2.5 创建类 DemoInfoServiceImpl

创建类 DemoInfoServiceImpl，代码如例 7-12 所示。

【例 7-12】 创建类 DemoInfoServiceImpl 的代码示例。

```
package com.bookcode.service.impl;
import com.bookcode.dao.DemoInfoRepository;
import com.bookcode.entity.DemoInfo;
import com.bookcode.service.DemoInfoService;
import javassist.NotFoundException;
import org.springframework.cache.annotation.CacheEvict;
import org.springframework.cache.annotation.CachePut;
import org.springframework.cache.annotation.Cacheable;
import org.springframework.stereotype.Service;
import javax.annotation.Resource;
import java.util.Optional;
@Service
public class DemoInfoServiceImpl implements DemoInfoService {
    //这里的单引号不能少，否则会报错，被识别的是一个对象
    public static final String CACHE_KEY = "'demoInfo'";
    @Resource
    private DemoInfoRepository demoInfoRepository;
    public static final String DEMO_CACHE_NAME = "demo";
    @CacheEvict(value=DEMO_CACHE_NAME,key=CACHE_KEY)
```

```
    @Override
    public DemoInfo save(DemoInfo demoInfo){
        return demoInfoRepository.save(demoInfo);
    }
    @Cacheable(value=DEMO_CACHE_NAME,key="'demoInfo_'+#id")
    @Override
    public Optional<DemoInfo> findById(Long id){
        System.err.println("没有走缓存! "+id);
        return demoInfoRepository.findById(id);
    }
    @CachePut(value = DEMO_CACHE_NAME,key = "'demoInfo_'+#updated.getId()")
    @Override
    public DemoInfo update(DemoInfo updated) throws NotFoundException {
        Optional<DemoInfo> opdemoInfo = demoInfoRepository.findById
       (updated.getId());
        if(opdemoInfo == null){
            throw new NotFoundException("No find");
        }
        DemoInfo demoInfo;
        opdemoInfo.get().setName(updated.getName());
        opdemoInfo.get().setPwd(updated.getPwd());
        demoInfo=opdemoInfo.get();
        return demoInfo;
    }
    @CacheEvict(value = DEMO_CACHE_NAME,key = "'demoInfo_'+#id")
    @Override
    public void delete(Long id) {
        System.out.println("模拟删除: "+id);
    }
}
```

7.2.6　创建类 DemoInfoController

创建类 DemoInfoController，代码如例 7-13 所示。

【例 7-13】 创建类 DemoInfoController 的代码示例。

```
package com.bookcode.controller;
import com.bookcode.entity.DemoInfo;
import com.bookcode.service.DemoInfoService;
import javassist.NotFoundException;
import org.springframework.web.bind.annotation.RequestMapping;
import org.springframework.web.bind.annotation.RestController;
import javax.annotation.Resource;
@RestController
public class DemoInfoController {
```

```
    @Resource
    private DemoInfoService demoInfoService;
    @RequestMapping("/test")
    public String test() {
        //存入两条数据
        DemoInfo demoInfo = new DemoInfo();
        demoInfo.setName("张三");
        demoInfo.setPwd("123456");
        DemoInfo demoInfo2 = demoInfoService.save(demoInfo);
        //不走缓存
        System.out.println(demoInfoService.findById(demoInfo2.getId()));
        //走缓存
        System.out.println(demoInfoService.findById(demoInfo2.getId()));
        demoInfo = new DemoInfo();
        demoInfo.setName("李四");
        demoInfo.setPwd("123456");
        DemoInfo demoInfo3 = demoInfoService.save(demoInfo);
        //不走缓存
        System.out.println(demoInfoService.findById(demoInfo3.getId()));
        //走缓存
        System.out.println(demoInfoService.findById(demoInfo3.getId()));
        System.out.println("============修改数据=====================");
        //修改数据
        DemoInfo updated = new DemoInfo();
        updated.setName("李四-updated");
        updated.setPwd("123456");
        updated.setId(demoInfo3.getId());
        try {
            System.out.println(demoInfoService.update(updated));
        }
    catch (NotFoundException e) {
            e.printStackTrace();
        }
        //不走缓存
        System.out.println(demoInfoService.findById(updated.getId()));
        //System.out.println(demoInfoService.findById(updated.getId()));
        return "ok";
    }
}
```

7.2.7 创建配置文件并运行程序

创建配置文件 application.yml，代码如例 7-6 所示。

运行程序后，在浏览器中输入 localhost:8080/test，在浏览器中输出 ok，结果如图 7-6

所示；同时，在控制台中的输出结果如图 7-7 所示。

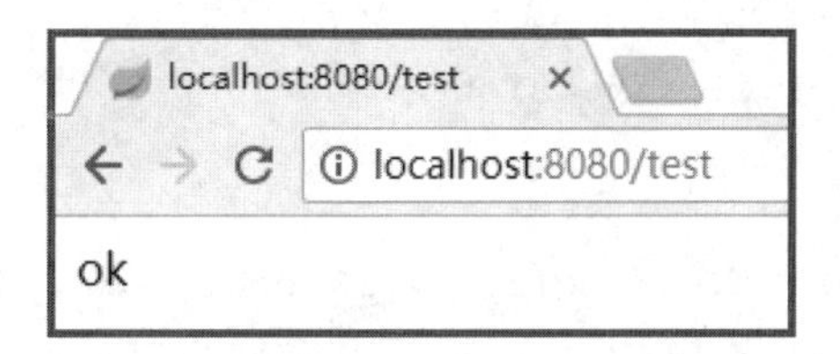

图 7-6　在浏览器中输入 localhost:8080/test 后输出 ok

```
Hibernate: select next_val as id_val from hibernate_sequence for update
Hibernate: update hibernate_sequence set next_val= ? where next_val=?
Hibernate: insert into demo_info (name, pwd, state, id) values (?, ?, ?, ?)
没有走缓存！3
Optional[DemoInfo [id=3, name=张三, pwd=123456, state=null]]
Optional[DemoInfo [id=3, name=张三, pwd=123456, state=null]]
Hibernate: select next_val as id_val from hibernate_sequence for update
Hibernate: update hibernate_sequence set next_val= ? where next_val=?
没有走缓存！4
Hibernate: insert into demo_info (name, pwd, state, id) values (?, ?, ?, ?)
Optional[DemoInfo [id=4, name=李四, pwd=123456, state=null]]
Optional[DemoInfo [id=4, name=李四, pwd=123456, state=null]]
============修改数据=====================
DemoInfo [id=4, name=李四-updated, pwd=123456, state=null]
Optional[DemoInfo [id=4, name=李四-updated, pwd=123456, state=null]]
```

图 7-7　在控制台中的输出结果

7.3　使用 Druid

视频讲解

Druid 是阿里巴巴开发的号称为监控而生的数据库连接池，它在功能、性能、扩展性方面都超过其他数据库连接池（如 DBCP、C3P0、BoneCP、Proxool、JBoss DataSource 等）。Druid 已经在阿里巴巴部署了许多应用，经过了生产环境大规模部署的严苛考验。Druid 不仅仅是一个数据库连接池，它的功能还包括监控和详细统计数据库访问性能，对数据库密码进行加密，提供不同的 LogFilter（支持 Common-Logging、Log4j 和 JdkLog 等），支持编写 JDBC 层的扩展插件。

7.3.1　添加依赖

在 pom.xml 文件中<dependencies>和</dependencies>之间添加依赖，代码如例 7-14 所示。

【例 7-14】 添加依赖的代码示例。

```
<dependency>
```

```
            <groupId>com.alibaba</groupId>
            <artifactId>druid</artifactId>
            <version>1.1.6</version>
</dependency>
<dependency>
            <groupId>org.springframework.boot</groupId>
            <artifactId>spring-boot-starter</artifactId>
</dependency>
<dependency>
            <groupId>org.springframework.boot</groupId>
            <artifactId>spring-boot-starter-web</artifactId>
</dependency>
<dependency>
            <groupId>mysql</groupId>
            <artifactId>mysql-connector-java</artifactId>
            <scope>runtime</scope>
</dependency>
<dependency>
            <groupId>javax.servlet</groupId>
            <artifactId>javax.servlet-api</artifactId>
            <version>3.1.0</version>
</dependency>
```

7.3.2 创建类 DruidStatViewServlet

在包 com.bookcode.servlet 中创建类 DruidStatViewServlet，代码如例 7-15 所示。

【例 7-15】 创建类 DruidStatViewServlet 的代码示例。

```
package com.bookcode.servlet;
import com.alibaba.druid.support.http.StatViewServlet;
import javax.servlet.annotation.WebInitParam;
import javax.servlet.annotation.WebServlet;
@WebServlet(urlPatterns="/druid/*",
        initParams={
                @WebInitParam(name="allow",value="192.168.1.72,127.0.0.1"),
  @WebInitParam(name="deny",value="192.168.1.73"),
                            //IP 黑名单(存在共同时， deny 优先于 allow)
                @WebInitParam(name="loginUsername",value="admin"), //用户名
                @WebInitParam(name="loginPassword",value="123456"),//密码
    @WebInitParam(name="resetEnable",value="false")
                            //禁用 HTML 页面上的 Reset All 功能
        }
)
public class DruidStatViewServlet extends StatViewServlet {
```

```
    private static final long serialVersionUID = 1L;
}
```

7.3.3　创建类 DruidStatFilter

在包 com.bookcode.servlet 中创建类 DruidStatFilter，代码如例 7-16 所示。

【例 7-16】 创建类 DruidStatFilter 的代码示例。

```
package com.bookcode.servlet;
import com.alibaba.druid.support.http.WebStatFilter;
import javax.servlet.annotation.WebFilter;
import javax.servlet.annotation.WebInitParam;
@WebFilter(filterName="druidWebStatFilter",urlPatterns="/*",
       initParams={
@WebInitParam(name="exclusions",value="*.js,*.gif,*.jpg,*.bmp,*.png,*
.c ss,*.ico,/druid/*")
       }
)
public class DruidStatFilter extends WebStatFilter {
}
```

7.3.4　修改入口类

修改入口类，代码如例 7-17 所示。

【例 7-17】 修改入口类的代码示例。

```
package com.bookcode;
import org.springframework.boot.SpringApplication;
import org.springframework.boot.autoconfigure.SpringBootApplication;
import org.springframework.boot.web.servlet.ServletComponentScan;
@SpringBootApplication
@ServletComponentScan
public class DemoApplication {
   public static void main(String[] args) {
      SpringApplication.run(DemoApplication.class, args);
   }
}
```

7.3.5　运行程序

运行程序后，在浏览器中输入 localhost:8080/druid/index.html，跳转到登录页面（localhost:8080/druid/login.html），结果如图 7-8 所示。输入正确的用户名（admin）、密码（123456）后，正常登录后的结果如图 7-9 所示。

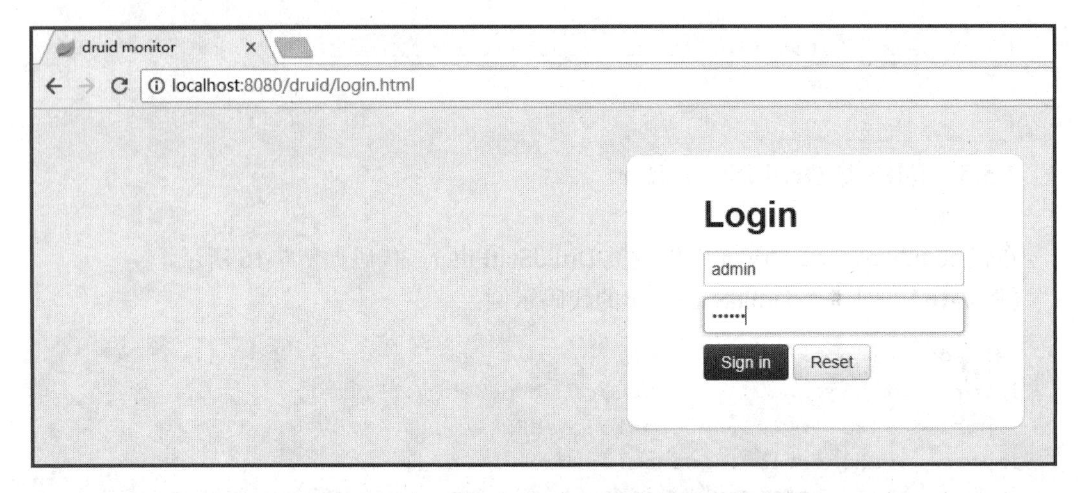

图 7-8 在浏览器中输入 localhost:8080/druid/index.html 的结果

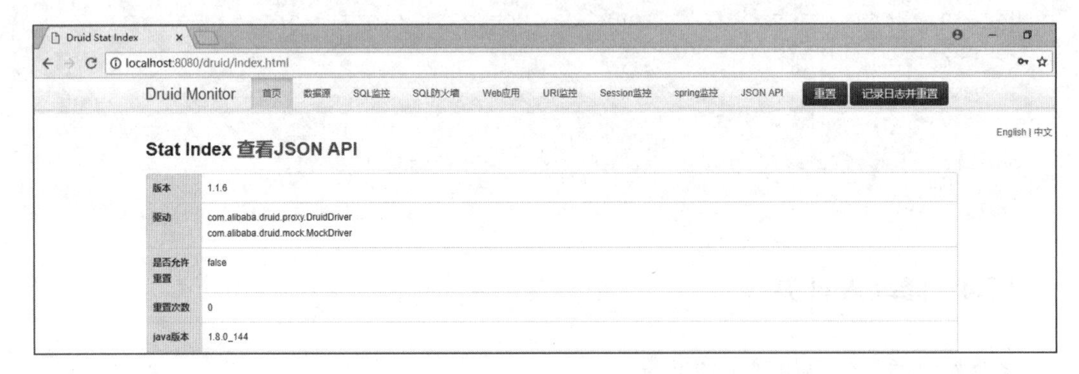

图 7-9 输入正确的用户名和密码后，正常登录后的结果

7.3.6 扩展程序并运行程序

在 com.bookcode.config 包中创建类 DruidConfiguration，代码如例 7-18 所示。

【例 7-18】 创建类 DruidConfiguration 的代码示例。

```
package com.bookcode.config;
import com.alibaba.druid.pool.DruidDataSource;
import com.alibaba.druid.support.http.StatViewServlet;
import com.alibaba.druid.support.http.WebStatFilter;
import org.springframework.beans.factory.annotation.Value;
import org.springframework.boot.web.servlet.FilterRegistrationBean;
import org.springframework.boot.web.servlet.ServletRegistrationBean;
import org.springframework.context.annotation.Bean;
import org.springframework.context.annotation.Configuration;
import javax.sql.DataSource;
import java.sql.SQLException;
@Configuration
```

```
public class DruidConfiguration {
    @Bean
    public ServletRegistrationBean DruidStatViewServle2() {
        ServletRegistrationBean servletRegistrationBean = new Servlet
        RegistrationBean(new
                StatViewServlet(), "/druid2/*");
        servletRegistrationBean.addInitParameter("allow", "127.0.0.1");
        servletRegistrationBean.addInitParameter("deny", "192.168.1.73");
        servletRegistrationBean.addInitParameter("loginUsername", "admin");
        servletRegistrationBean.addInitParameter("loginPassword", "123456");
        servletRegistrationBean.addInitParameter("resetEnable", "true");
        return servletRegistrationBean;
    }
    @Bean
    public FilterRegistrationBean druidStatFilter2(){
  FilterRegistrationBean filterRegistrationBean = new FilterRegistrationBean
  (new WebStatFilter());
        filterRegistrationBean.addUrlPatterns("/*");
    filterRegistrationBean.addInitParameter("exclusions","*.js,*.gif,
    *.jpg,*.png,*.css,*.ico,/druid2/*");
        return filterRegistrationBean;
    }
    @Bean
  public  DataSource  druidDataSource(@Value("${spring.datasource.driver-
  class-name}") String driver,
                                @Value("${spring.datasource.url}") String url,
                     @Value("${spring.datasource.username}") String username,
                     @Value("${spring.datasource.password}") String password){
        DruidDataSource druidDataSource = new DruidDataSource();
        druidDataSource.setDriverClassName(driver);
        druidDataSource.setUrl(url);
        druidDataSource.setUsername(username);
        druidDataSource.setPassword(password);
      System.out.println("DruidConfiguration.druidDataSource(),url=
      "+url+",username="+username+",password="+password+".");
        try {
            druidDataSource.setFilters("stat, wall");
        }
        catch (SQLException e) {
            e.printStackTrace();
        }
        return druidDataSource;
    }
}
```

例 7-18 在 DruidConfiguration 中加入了一个方法 druidDataSource()来进行进行数据源的注入代码，还需要增加一个配置文件 application.yml，代码如例 7-6 所示。一般来说，通过配置文件的方式注入比较好，如果有特殊需求，也可以以编程方式进行注入。其他内容保持不变并运行程序，再在浏览器中输入 localhost:8080/druid2/index.html，自动跳转到登录页面（localhost:8080/druid/login.html）。

7.4 使用表单验证

视频讲解

表单验证即校验用户提交的数据的合理性，例如，密码长度是否大于 6 位。Spring Boot 可以使用注解 @Valid 进行表单验证。常见的注解还包括：

（1）@Null 限制只能为 null。

（2）@NotNull 限制必须不为 null。

（3）@AssertFalse 限制必须为 false。

（4）@AssertTrue 限制必须为 true。

（5）@DecimalMax(value)限制必须为一个不大于指定值 value 的数。

（6）@DecimalMin(value)限制必须为一个不小于指定值 value 的数。

（7）@Digits(integer,fraction)限制必须为一个小数，且整数部分的位数不能超过 integer，小数部分的位数不能超过 fraction。

（8）@Future 限制必须为一个将来的日期。

（9）@Max(value)限制必须为一个不大于指定值 value 的数。

（10）@Min(value)限制必须为一个不小于指定值 value 的数。

（11）@Past 限制必须为一个过去的日期。

（12）@Pattern(value)限制必须符合指定的正则表达式 value。

（13）@Size(min, max)限制字符长度必须为 min～max。

（14）@NotEmpty 限制注解的元素值不为 null 且不为空（字符串长度不为 0、集合大小不为 0）。

（15）@NotBlank 限制注解的元素值不为空（不为 null、去除首位空格后长度为 0），不同于@NotEmpty，@NotBlank 只应用于字符串且在比较时会去除字符串的空格。

（16）@Email 限制注解的元素值是 Email，也可以通过正则表达式和 flag 指定自定义的 Email 格式。

7.4.1 添加依赖

在 pom.xml 文件中<dependencies>和</dependencies>之间添加依赖，代码如例 7-19 所示。

【例 7-19】 添加依赖的代码示例。

```
<dependency>
        <groupId>org.springframework.boot</groupId>
```

```
        <artifactId>spring-boot-starter-web</artifactId>
</dependency>
<dependency>
        <groupId>mysql</groupId>
        <artifactId>mysql-connector-java</artifactId>
</dependency>
<dependency>
        <groupId>org.springframework.boot</groupId>
        <artifactId>spring-boot-starter-data-jpa</artifactId>
</dependency>
```

7.4.2　创建类 Student

创建类 Student，代码如例 7-20 所示。

【例 7-20】 创建类 Student 的代码示例。

```
package com.bookcode.entity;
import javax.persistence.*;
import javax.validation.constraints.Min;
import javax.validation.constraints.NotEmpty;
import javax.validation.constraints.NotNull;
@Entity
@Table(name="t_student")
public class Student {
   @Id
   @GeneratedValue
   private Integer id;
   @NotEmpty(message="姓名不能为空！")
   @Column(length=50)
   private String name;
   @NotNull(message="年龄不能为空！")
   @Min(value=18,message="年龄必须大于 18 岁")
   private Integer age;
   public Integer getId() {
      return id;    }
   public void setId(Integer id) {
      this.id = id;    }
   public String getName() {
      return name;    }
   public void setName(String name) {
      this.name = name;    }
   public Integer getAge() {
      return age;    }
```

```
    public void setAge(Integer age) {
        this.age = age;    }
}
```

7.4.3 创建接口 StudentDao

创建接口 StudentDao，代码如例 7-21 所示。

【例 7-21】 创建接口 StudentDao 的代码示例。

```
package com.bookcode.dao;
import com.bookcode.entity.Student;
import org.springframework.data.jpa.repository.JpaRepository;
public interface StudentDao extends JpaRepository<Student,Integer> {
}
```

7.4.4 创建接口 StudentService

创建接口 StudentService，代码如例 7-22 所示。

【例 7-22】 创建接口 StudentService 的代码示例。

```
package com.bookcode.service;
import com.bookcode.entity.Student;
public interface StudentService {
    //添加学生信息到数据库
    public void add(Student student);
}
```

7.4.5 创建类 StudentServiceImpl

创建类 StudentServiceImpl，代码如例 7-23 所示。

【例 7-23】 创建类 StudentServiceImpl 的代码示例。

```
package com.bookcode.service.impl;
import com.bookcode.dao.StudentDao;
import com.bookcode.entity.Student;
import com.bookcode.service.StudentService;
import org.springframework.stereotype.Service;
import javax.annotation.Resource;
@Service("studentService")
public class StudentServiceImpl implements StudentService {
    @Resource
    private StudentDao studentDao;
    @Override
    public void add(Student student) {
```

```
        studentDao.save(student);
    }
}
```

7.4.6　创建类 StudentController

创建类 StudentController，代码如例 7-24 所示。

【例 7-24】 创建类 StudentController 的代码示例。

```
package com.bookcode.controller;
import com.bookcode.entity.Student;
import com.bookcode.service.StudentService;
import org.springframework.validation.BindingResult;
import org.springframework.web.bind.annotation.RequestMapping;
import org.springframework.web.bind.annotation.RestController;
import javax.annotation.Resource;
import javax.validation.Valid;
@RestController
@RequestMapping("/student")
public class StudentController {
    @Resource
    private StudentService studentService;
    @RequestMapping("/add")
    public String add(@Valid Student student, BindingResult bindingResult){
        if(bindingResult.hasErrors()){
            return bindingResult.getFieldError().getDefaultMessage();
        }
        else{
            studentService.add(student);
            return "添加成功";
        }
    }
}
```

7.4.7　创建文件 studentAdd.html

在 resources/static 目录下创建文件 studentAdd.html，代码如例 7-25 所示。

【例 7-25】 创建文件 studentAdd.html 的代码示例。

```
<!DOCTYPE html>
<html>
<head>
    <meta charset="UTF-8"/>
    <title>学生信息添加页面</title>
```

```
        <script  src="http://www.java1234.com/jquery-easyui-1.3.3/jquery.min.
        js"></script>
        <script type="text/javascript">
            function submitData(){
                $.post("/student/add",{name:$("#name").val(),age:$("#age").val()},
                    function(result){
                        alert(result);
                    }
                );
            }
        </script>
    </head>
    <body>
    姓名：<input type="text" id="name" name="name"/><br/>
    年龄：<input type="text" id="age" name="age"/><br/>
    <input type="button" onclick="submitData()" value="提交"/>
    </body>
    </html>
```

7.4.8 创建配置文件并运行程序

创建配置文件 application.yml，代码如例 7-6 所示。

运行程序后，在浏览器中输入 localhost:8080/studentAdd.html，假如输入的学生信息符合在实体类 Student 中设置的条件，结果如图 7-10 所示。假如输入的学生信息不符合在实体类 Student 中设置的条件，结果如图 7-11 所示。此时需要重新输入学生信息。学生信息输入成功后，在工具 Navicat for MySQL 中的结果如图 7-12 所示。

图 7-10 输入的学生信息符合在实体类 Student 中设置的条件的结果

图 7-11 输入的学生信息不符合在实体类 Student 中设置的条件的结果

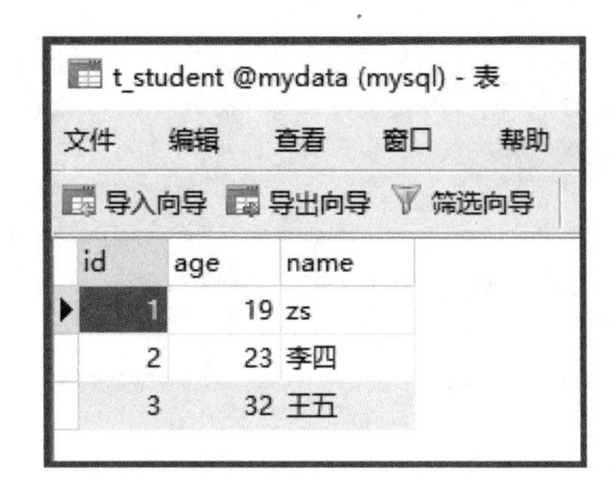

图 7-12　信息输入成功后在工具 Navicat for MySQL 中的结果

7.5　整合 MyBatis 访问数据库

视频讲解

MyBatis 源自 Apache 的开源项目 iBatis，2010 年改名为 MyBatis。iBatis 是 internet 和 abatis 的组合，iBatis 包括 SQL Maps 和 Data Access Objects（DAO）。MyBatis 是一款优秀的持久层框架，它支持定制化 SQL、存储过程以及高级映射。MyBatis 避免了几乎所有的 JDBC 代码和手动设置参数以及获取结果集。MyBatis 可使用简单的 XML 或注解将接口和普通 Java 对象（Plain Old Java Objects，POJO）映射成数据库中的记录。

每个 MyBatis 应用程序都要使用 SqlSessionFactory 实例，SqlSessionFactory 实例可以通过 SqlSessionFactoryBuilder 获得。SqlSessionFactoryBuilder 可以由 XML 配置文件或者预定义的配置类的实例获得。

7.5.1　添加依赖

在 pom.xml 文件中<dependencies>和</dependencies>之间添加 Web、MyBatis、MySQL 依赖，代码如例 7-26 所示。代码中最后一对<dependency>和</dependency>之间的 test 依赖是自动添加的，不是手动添加的结果。后面章节中添加依赖的代码中若有 test 依赖也是如此，不再一一说明。

【例 7-26】 添加依赖的代码示例。

```
<dependency>
        <groupId>org.springframework.boot</groupId>
        <artifactId>spring-boot-starter-web</artifactId>
</dependency>
<dependency>
        <groupId>org.mybatis.spring.boot</groupId>
        <artifactId>mybatis-spring-boot-starter</artifactId>
        <version>1.2.0</version>
</dependency>
<dependency>
```

```
            <groupId>mysql</groupId>
            <artifactId>mysql-connector-java</artifactId>
</dependency>
<dependency>
            <groupId>org.springframework.boot</groupId>
            <artifactId>spring-boot-starter-test</artifactId>
            <scope>test</scope>
</dependency>
```

7.5.2 创建类 City

创建类 City，代码如例 7-27 所示。

【例 7-27】 创建类 City 的代码示例。

```
package com.bookcode.entity;
public class City {
    private Long id;               //城市编号
    private Long provinceId;       //省份编号
    private String cityName;
    private String description;
    public Long getId() {
        return id;
    }
    public void setId(Long id) {
        this.id = id;
    }
    public Long getProvinceId() {
        return provinceId;
    }
    public void setProvinceId(Long provinceId) {
        this.provinceId = provinceId;
    }
    public String getCityName() {
        return cityName;
    }
    public void setCityName(String cityName) {
        this.cityName = cityName;
    }
    public String getDescription() {
        return description;
    }
    public void setDescription(String description) {
        this.description = description;
    }
}
```

7.5.3　创建接口 CityDao

创建接口 CityDao，代码如例 7-28 所示。

【例 7-28】 创建接口 CityDao 的代码示例。

```
package com.bookcode.dao;
import com.bookcode.entity.City;
import org.apache.ibatis.annotations.*;
import java.util.List;
@Mapper         //标志为 MyBatis 的 Mapper
public interface CityDao {
    @Results({
           @Result(property = "id", column = "id"),
           @Result(property = "provinceId", column = "provinceId"),
           @Result(property = "cityName", column = "cityName"),
           @Result(property = "description", column = "description"),
    })
    @Select("SELECT * FROM city WHERE cityName = #{cityName}")
    List<City> findByName(String cityName);
}
```

7.5.4　创建接口 CityService

创建接口 CityService，代码如例 7-29 所示。

【例 7-29】 创建接口 CityService 的代码示例。

```
package com.bookcode.service;
import com.bookcode.entity.City;
import java.util.List;
public interface CityService {
   List<City> findCityByName(String cityName);
}
```

7.5.5　创建类 CityServiceImpl

创建类 CityServiceImpl，代码如例 7-30 所示。

【例 7-30】 创建类 CityServiceImpl 的代码示例。

```
package com.bookcode.service.impl;
import com.bookcode.dao.CityDao;
import com.bookcode.entity.City;
import com.bookcode.service.CityService;
import org.springframework.stereotype.Service;
import javax.annotation.Resource;
import java.util.List;
```

```
@Service
public class CityServiceImpl implements CityService {
    @Resource
    private CityDao cityDao;
    public List <City> findCityByName(String cityName) {
        return cityDao.findByName(cityName);
    }
}
```

7.5.6 创建类 CityController

创建类 CityController，代码如例 7-31 所示。

【例 7-31】 创建类 CityController 的代码示例。

```
package com.bookcode.controller;
import com.bookcode.entity.City;
import com.bookcode.service.CityService;
import org.springframework.web.bind.annotation.RequestMapping;
import org.springframework.web.bind.annotation.RequestMethod;
import org.springframework.web.bind.annotation.RequestParam;
import org.springframework.web.bind.annotation.RestController;
import javax.annotation.Resource;
import java.util.List;
@RestController
public class CityController {
    @Resource
    private CityService cityService;
    @RequestMapping(value = "/api/city", method = RequestMethod.GET)
    public List<City> findOneCity(@RequestParam(value="cityName", required=
    true) String cityName) {
        return cityService.findCityByName(cityName);
    }
}
```

7.5.7 修改配置文件 application.properties

配置文件 application.properties 的代码如例 7-32 所示。

【例 7-32】 配置文件 application.properties 的代码示例。

```
spring.datasource.url=jdbc:mysql://localhost:3306/mytest?useUnicode=tru
e&characterEncoding=utf8
spring.datasource.username=root
spring.datasource.password=sa
spring.datasource.driver-class-name=com.mysql.jdbc.Driver
```

7.5.8　运行程序

运行程序后，在浏览器中输入 localhost:8080/api/city?cityName=xz，结果如图 7-13 所示。

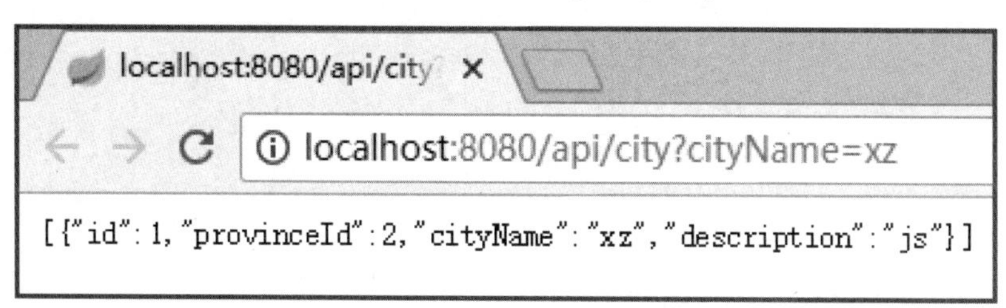

图 7-13　在浏览器中输入 localhost:8080/api/city?cityName=xz 后的结果

7.6　整合 Spring Batch 和 Quartz

视频讲解

Spring Batch 是一个完善的轻量级批处理框架，旨在帮助企业建立健壮、高效的批处理应用。Spring Batch 是使用 Java 语言并基于 Spring 框架为基础开发的。Spring Batch 提供了大量可重用的组件，包括日志、追踪、事务、任务作业统计、任务重启、跳过、资源管理。对于大数据量和高性能的批处理任务，Spring Batch 同样提供了高级功能和特性来支持，如分区、远程功能。Spring Batch 不是调度框架，需要和调度框架（如 Quartz 等）合作来构建完成批处理任务。

Quartz 是一个用 Java 编写的开源作业调度框架，为在 Java 应用程序中进行作业调度提供了简单而强大的机制。Quartz 可以与 Java EE、Java SE 应用程序相结合，也可以单独使用。Quartz 允许程序开发人员根据时间的间隔来调度作业。Quartz 实现了作业和触发器之间的多对多关系，还能把多个作业与不同的触发器关联。为确保可伸缩性，Quartz 采用了基于多线程的架构。

7.6.1　添加依赖

在 pom.xml 文件中<dependencies>和</dependencies>之间添加依赖，代码如例 7-33 所示。

【例 7-33】 添加依赖的代码示例。

```
<dependency>
        <groupId>org.springframework.boot</groupId>
        <artifactId>spring-boot-starter</artifactId>
</dependency>
<dependency>
        <groupId>org.springframework.boot</groupId>
        <artifactId>spring-boot-starter-test</artifactId>
        <scope>test</scope>
</dependency>
<dependency>
```

```
            <groupId>org.springframework.boot</groupId>
            <artifactId>spring-boot-starter-batch</artifactId>
</dependency>
<dependency>
            <groupId>com.h2database</groupId>
            <artifactId>h2</artifactId>
            <scope>runtime</scope>
</dependency>
```

7.6.2 创建类 MyTaskOne

创建类 MyTaskOne，代码如例 7-34 所示。

【例 7-34】 创建类 MyTaskOne 的代码示例。

```
package com.bookcode.batchandquartz.task;
import org.springframework.batch.core.StepContribution;
import org.springframework.batch.core.scope.context.ChunkContext;
import org.springframework.batch.core.step.tasklet.Tasklet;
import org.springframework.batch.repeat.RepeatStatus;
public class MyTaskOne implements Tasklet {
    public RepeatStatus execute(StepContribution contribution, ChunkContext
    chunkContext) throws Exception
    {
        System.out.println("MyTaskOne start..");
        System.out.println("My Code of TaskOne..");
        System.out.println("MyTaskOne done..");
        return RepeatStatus.FINISHED;
    }
}
```

7.6.3 创建类 MyTaskTwo

创建类 MyTaskTwo，代码如例 7-35 所示。

【例 7-35】 创建类 MyTaskTwo 的代码示例。

```
package com.bookcode.batchandquartz.task;
import org.springframework.batch.core.StepContribution;
import org.springframework.batch.core.scope.context.ChunkContext;
import org.springframework.batch.core.step.tasklet.Tasklet;
import org.springframework.batch.repeat.RepeatStatus;
public class MyTaskTwo implements Tasklet {
    public RepeatStatus execute(StepContribution contribution, ChunkContext
    chunkContext) throws Exception
    {
```

```
        System.out.println("MyTaskTwo start..");
        System.out.println("My code of  TaskTwo ..");
          System.out.println("MyTaskTwo done..");
        return RepeatStatus.FINISHED;
    }
}
```

7.6.4　创建类 BatchConfig

创建类 BatchConfig，代码如例 7-36 所示。

【例 7-36】 创建类 BatchConfig 的代码示例。

```
package com.bookcode.batchandquartz.config;
import com.bookcode.batchandquartz.task.MyTaskOne;
import com.bookcode.batchandquartz.task.MyTaskTwo;
import org.springframework.batch.core.Job;
import org.springframework.batch.core.Step;
import org.springframework.batch.core.configuration.annotation.
Enable BatchProcessing;
import org.springframework.batch.core.configuration.annotation.
JobBuilderFactory;
import org.springframework.batch.core.configuration.annotation.
StepBuilderFactory;
import org.springframework.batch.core.launch.support.RunIdIncrementer;
import org.springframework.beans.factory.annotation.Autowired;
import org.springframework.context.annotation.Bean;
import org.springframework.context.annotation.Configuration;
@Configuration
@EnableBatchProcessing
public class BatchConfig {
   @Autowired
   private JobBuilderFactory jobs;
   @Autowired
   private StepBuilderFactory steps;
   @Bean
   public Step stepOne(){
      return steps.get("stepOne")
            .tasklet(new MyTaskOne())
            .build();
   }
   @Bean
   public Step stepTwo(){
      return steps.get("stepTwo")
            .tasklet(new MyTaskTwo())
            .build();
```

```
    }
    @Bean
    public Job demoJob(){
        return jobs.get("demoJob")
                .incrementer(new RunIdIncrementer())
                .start(stepOne())
                .next(stepTwo())
                .build();
    }
}
```

7.6.5 修改入口类

修改入口类，代码如例 7-37 所示。

【例 7-37】 修改入口类的代码示例。

```
package com.bookcode.batchandquartz;
import org.springframework.batch.core.Job;
import org.springframework.batch.core.JobParameters;
import org.springframework.batch.core.JobParametersBuilder;
import org.springframework.batch.core.launch.JobLauncher;
import org.springframework.beans.factory.annotation.Autowired;
import org.springframework.boot.CommandLineRunner;
import org.springframework.boot.SpringApplication;
import org.springframework.boot.autoconfigure.SpringBootApplication;
@SpringBootApplication
public class BatchandquartzApplication implements CommandLineRunner {
    @Autowired
    JobLauncher jobLauncher;
    @Autowired
    Job job;
    public static void main(String[] args) {
        SpringApplication.run(BatchandquartzApplication.class, args);
    }
    @Override
    public void run(String... args) throws Exception
    {
        JobParameters params = new JobParametersBuilder()
                .addString("JobID", String.valueOf(System.currentTimeMillis()))
                .toJobParameters();
        jobLauncher.run(job, params);
    }
}
```

7.6.6 运行程序

运行程序后，控制台中的输出结果如图 7-14 所示。

```
MyTaskOne start..
My Code of TaskOne..
MyTaskOne done..
2018-10-09 20:30:32.878
MyTaskTwo start..
My code of  TaskTwo ..
MyTaskTwo done..
```

图 7-14 控制台中的输出结果

7.6.7 增加依赖

在 pom.xml 文件中<dependencies>和</dependencies>之间增加 Quartz 依赖，代码如例 7-38 所示。

【例 7-38】 增加 Quartz 依赖的代码示例。

```
<dependency>
        <groupId>org.springframework.boot</groupId>
        <artifactId>spring-boot-starter-quartz</artifactId>
</dependency>
```

7.6.8 修改类 BatchConfig

修改类 BatchConfig，代码如例 7-39 所示。

【例 7-39】 修改类 BatchConfig 的代码示例。

```
package com.bookcode.batchandquartz.config;
import com.bookcode.batchandquartz.task.MyTaskOne;
import com.bookcode.batchandquartz.task.MyTaskTwo;
import org.springframework.batch.core.Job;
import org.springframework.batch.core.Step;
import org.springframework.batch.core.configuration.annotation.Enable
BatchProcessing;
import org.springframework.batch.core.configuration.annotation.
JobBuilderFactory;
import org.springframework.batch.core.configuration.annotation.
StepBuilderFactory;
```

```
import org.springframework.beans.factory.annotation.Autowired;
import org.springframework.context.annotation.Bean;
import org.springframework.context.annotation.Configuration;
@Configuration
@EnableBatchProcessing
public class BatchConfig {
    @Autowired
    private JobBuilderFactory jobs;
    @Autowired
    private StepBuilderFactory steps;
    @Bean
    public Step stepOne(){
        return steps.get("stepOne")
                .tasklet(new MyTaskOne())
                .build();
    }
    @Bean
    public Step stepTwo(){
        return steps.get("stepTwo")
                .tasklet(new MyTaskTwo())
                .build();
    }
    @Bean(name="demoJobOne")
    public Job demoJobOne(){
        return jobs.get("demoJobOne")
                .start(stepOne())
                .next(stepTwo())
                .build();
    }
    @Bean(name="demoJobTwo")
    public Job demoJobTwo(){
        return jobs.get("demoJobTwo")
                .flow(stepOne())
                .build()
                .build();
    }
}
```

7.6.9 创建类 CustomQuartzJob

创建 Quartz 作业运行器类 CustomQuartzJob，代码如例 7-40 所示。

【例 7-40】 创建类 CustomQuartzJob 的代码示例。

```
package com.bookcode.batchandquartz.jobs;
import org.quartz.JobExecutionContext;
```

```
import org.quartz.JobExecutionException;
import org.springframework.batch.core.Job;
import org.springframework.batch.core.JobParameters;
import org.springframework.batch.core.JobParametersBuilder;
import org.springframework.batch.core.configuration.JobLocator;
import org.springframework.batch.core.launch.JobLauncher;
import org.springframework.scheduling.quartz.QuartzJobBean;
public class CustomQuartzJob extends QuartzJobBean {
    private String jobName;
    private JobLauncher jobLauncher;
    private JobLocator jobLocator;
    public String getJobName() {
        return jobName;
    }
    public void setJobName(String jobName) {
        this.jobName = jobName;
    }
    public JobLauncher getJobLauncher() {
        return jobLauncher;
    }
    public void setJobLauncher(JobLauncher jobLauncher) {
        this.jobLauncher = jobLauncher;
    }
    public JobLocator getJobLocator() {
        return jobLocator;
    }
    public void setJobLocator(JobLocator jobLocator) {
        this.jobLocator = jobLocator;
    }
    @Override
    protected void executeInternal(JobExecutionContext context) throws
    JobExecutionException
    {
        try
        {
            Job job = jobLocator.getJob(jobName);
            JobParameters params = new JobParametersBuilder()
                    .addString("JobID", String.valueOf(System.currentTimeMillis()))
                    .toJobParameters();
            jobLauncher.run(job, params);
        }
        catch (Exception e)
        {
            e.printStackTrace();
```

```
        }
    }
}
```

7.6.10 创建类 QuartzConfig

创建类 QuartzConfig，代码如例 7-41 所示。

【例 7-41】 创建类 QuartzConfig 的代码示例。

```
package com.bookcode.batchandquartz.config;
import java.io.IOException;
import java.util.Properties;
import com.bookcode.batchandquartz.jobs.CustomQuartzJob;
import org.quartz.JobBuilder;
import org.quartz.JobDataMap;
import org.quartz.JobDetail;
import org.quartz.SimpleScheduleBuilder;
import org.quartz.Trigger;
import org.quartz.TriggerBuilder;
import org.springframework.batch.core.configuration.JobLocator;
import org.springframework.batch.core.configuration.JobRegistry;
import org.springframework.batch.core.configuration.support.
JobRegistry BeanPostProcessor;
import org.springframework.batch.core.launch.JobLauncher;
import org.springframework.beans.factory.annotation.Autowired;
import org.springframework.beans.factory.config.PropertiesFactoryBean;
import org.springframework.context.annotation.Bean;
import org.springframework.context.annotation.Configuration;
import org.springframework.core.io.ClassPathResource;
import org.springframework.scheduling.quartz.SchedulerFactoryBean;
@Configuration
public class QuartzConfig
{
    @Autowired
    private JobLauncher jobLauncher;
    @Autowired
    private JobLocator jobLocator;
    @Bean
    public JobRegistryBeanPostProcessor jobRegistryBeanPostProcessor
    (JobRegistry jobRegistry) {
        JobRegistryBeanPostProcessor jobRegistryBeanPostProcessor = new
        JobRegistryBeanPostProcessor();
        jobRegistryBeanPostProcessor.setJobRegistry(jobRegistry);
        return jobRegistryBeanPostProcessor;
    }
```

```
@Bean
public JobDetail jobOneDetail() {
    //Set Job data map
    JobDataMap jobDataMap = new JobDataMap();
    jobDataMap.put("jobName", "demoJobOne");
    jobDataMap.put("jobLauncher", jobLauncher);
    jobDataMap.put("jobLocator", jobLocator);
    return JobBuilder.newJob(CustomQuartzJob.class)
            .withIdentity("demoJobOne")
            .setJobData(jobDataMap)
            .storeDurably()
            .build();
}
@Bean
public JobDetail jobTwoDetail() {
    JobDataMap jobDataMap = new JobDataMap();
    jobDataMap.put("jobName", "demoJobTwo");
    jobDataMap.put("jobLauncher", jobLauncher);
    jobDataMap.put("jobLocator", jobLocator);
    return JobBuilder.newJob(CustomQuartzJob.class)
            .withIdentity("demoJobTwo")
            .setJobData(jobDataMap)
            .storeDurably()
            .build();
}
@Bean
public Trigger jobOneTrigger()
{
    SimpleScheduleBuilder scheduleBuilder = SimpleScheduleBuilder
            .simpleSchedule()
            .withIntervalInSeconds(10)
            .repeatForever();
    return TriggerBuilder
            .newTrigger()
            .forJob(jobOneDetail())
            .withIdentity("jobOneTrigger")
            .withSchedule(scheduleBuilder)
            .build();
}
@Bean
public Trigger jobTwoTrigger()
{
    SimpleScheduleBuilder scheduleBuilder = SimpleScheduleBuilder
            .simpleSchedule()
            .withIntervalInSeconds(20)
```

```
                .repeatForever();
        return TriggerBuilder
                .newTrigger()
                .forJob(jobTwoDetail())
                .withIdentity("jobTwoTrigger")
                .withSchedule(scheduleBuilder)
                .build();
    }
    @Bean
    public SchedulerFactoryBean schedulerFactoryBean() throws IOException
    {
        SchedulerFactoryBean scheduler = new SchedulerFactoryBean();
        scheduler.setTriggers(jobOneTrigger(), jobTwoTrigger());
        scheduler.setQuartzProperties(quartzProperties());
        scheduler.setJobDetails(jobOneDetail(), jobTwoDetail());
        return scheduler;
    }
    @Bean
    public Properties quartzProperties() throws IOException
    {
        PropertiesFactoryBean propertiesFactoryBean = new Properties
        FactoryBean();
        propertiesFactoryBean.setLocation(new ClassPathResource("/quartz.
        properties"));
        propertiesFactoryBean.afterPropertiesSet();
        return propertiesFactoryBean.getObject();
    }
}
```

7.6.11 创建文件 quartz.properties 和 application.properties

在 resources 目录下创建文件 quartz.properties，代码如例 7-42 所示。

【例 7-42】 创建文件 quartz.properties 的代码示例。

```
#scheduler name will be "MyScheduler"
org.quartz.scheduler.instanceName = MyScheduler
#maximum of 3 jobs can be run simultaneously
org.quartz.threadPool.threadCount = 3
#All of Quartz data is held in memory (rather than in a database)
org.quartz.jobStore.class = org.quartz.simpl.RAMJobStore
```

在 resources 目录下创建文件 application.properties，代码如例 7-43 所示。

【例 7-43】 创建文件 application.properties 的代码示例。

```
spring.batch.job.enabled=false
spring.h2.console.enabled=true
```

7.6.12　修改入口类

修改入口类，代码如例 7-44 所示。

【例 7-44】 修改入口类的代码示例。

```
package com.bookcode.batchandquartz;
import org.springframework.boot.SpringApplication;
import org.springframework.boot.autoconfigure.SpringBootApplication;
@SpringBootApplication
public class BatchandquartzApplication   {
    public static void main(String[] args) {
        SpringApplication.run(BatchandquartzApplication.class, args);
    }
}
```

7.6.13　运行程序

运行程序后，控制台中的输出结果依次如图 7-15～图 7-18 所示。

```
Scheduler class: 'org.quartz.core.QuartzScheduler' - running locally.
NOT STARTED.
Currently in standby mode.
Number of jobs executed: 0
Using thread pool 'org.quartz.simpl.SimpleThreadPool' - with 3 threads.
Using job-store 'org.quartz.simpl.RAMJobStore' - which does not support persistence. and is not clustered.
```

图 7-15　控制台中的输出结果（第一部分）

```
MyTaskOne start..
My Code of TaskOne..
MyTaskOne done..
MyTaskOne start..
My Code of TaskOne..
MyTaskOne done..
MyTaskOne start..
```

图 7-16　控制台中的输出结果（第二部分）

```
MyTaskTwo start..
My code of  TaskTwo ..
MyTaskTwo start..
My code of  TaskTwo ..
MyTaskTwo done..
MyTaskTwo done..
```

图 7-17　控制台中的输出结果（第三部分）

```
MyTaskOne start..
My Code of TaskOne..
MyTaskOne done..
2018-10-09 21:03:18.705  INFO
MyTaskTwo start..
My code of  TaskTwo ..
MyTaskTwo done..
```

图 7-18　控制台中的输出结果（第四部分）

习题 7

简答题

1．简述对声明式事务的理解。

2．简述对数据缓存的理解。

3．简述对 Druid 的理解。

4．简述用于表单验证的常用注解的作用。

5．简述对 MyBatis 的理解。

6．简述对 Spring Batch 的理解。

7．简述对 Quartz 的理解。

实验题

1．实现声明式事务。

2．实现数据缓存。

3．实现对 Druid 的使用。

4．实现表单验证。

5．实现通过整合 MyBatis 来访问数据库。

6．实现对 Spring Batch 和 Quartz 的整合。

第 8 章

Spring Boot 的文件应用

本章介绍如何实现文件上传、如何实现文件下载、如何实现图片文件上传和显示、如何访问 HDFS、如何利用 Elasticsearch 实现全文搜索、如何实现邮件发送、如何实现用 REST Docs 创建 API 文档等内容。

8.1 文件上传

视频讲解

Web 应用项目中经常会有上传和下载的需求，本节结合实例介绍如何用 Spring Boot 实现文件的上传。

8.1.1 添加依赖

在 pom.xml 文件中<dependencies>和</dependencies>之间添加依赖，代码如例 8-1 所示。

【例 8-1】 添加依赖的代码示例。

```
<dependency>
          <groupId>org.springframework.boot</groupId>
          <artifactId>spring-boot-starter-web</artifactId>
</dependency>
<dependency>
          <groupId>org.springframework.boot</groupId>
          <artifactId>spring-boot-starter-thymeleaf</artifactId>
</dependency>
```

8.1.2 创建类 FileUploadController

创建类 FileUploadController，代码如例 8-2 所示。

【例 8-2】 创建类 FileUploadController 的代码示例。

```
package com.bookcode.controller;
import org.springframework.stereotype.Controller;
import org.springframework.web.bind.annotation.RequestMapping;
import org.springframework.web.bind.annotation.RequestMethod;
import org.springframework.web.bind.annotation.RequestParam;
import org.springframework.web.bind.annotation.ResponseBody;
import org.springframework.web.multipart.MultipartFile;
import org.springframework.web.multipart.MultipartHttpServletRequest;
import javax.servlet.http.HttpServletRequest;
import java.io.*;
import java.util.List;
@Controller
public class FileUploadController {
    @RequestMapping("/file")
    public String file() {
        return "/file";    }
    @RequestMapping("/upload")
    @ResponseBody
    public String handleFileUpload(@RequestParam("file")MultipartFile file) {
        if(!file.isEmpty()) {
            try {
    BufferedOutputStream out=new BufferedOutputStream(new FileOutputStream
    (new File(file.getOriginalFilename())));
                out.write(file.getBytes());
                out.flush();
                out.close();
            }
              catch(FileNotFoundException e) {
                e.printStackTrace();
                return "上传失败," + e.getMessage();
            }
              catch(IOException e) {
                e.printStackTrace();
                return "上传失败," + e.getMessage();
            }
            return "上传成功";
        }
          else {
            return "上传失败，因为文件是空的.";
        }
```

```
    }
    @RequestMapping("/multifile")
    public String multifile() {
        return "/multifile";
        }
    @RequestMapping(value="/batch/upload", method= RequestMethod.POST)
    public@ResponseBody
    String handleFileUpload(HttpServletRequest request){
        List<MultipartFile> files = ((MultipartHttpServletRequest)request).
        getFiles("file");
        MultipartFile file = null;
        BufferedOutputStream stream = null;
        for(int i =0; i< files.size(); ++i) {
            file = files.get(i);
            if(!file.isEmpty()) {
                try {
                    byte[] bytes = file.getBytes();
                    stream =new BufferedOutputStream(new FileOutputStream(new
                    File(file.getOriginalFilename())));
                    stream.write(bytes);
                    stream.close();
                }
            catch(Exception e) {
            stream = null;
            return "You failed to upload " + i + " => " + e.getMessage();
                }
            }
               else {
                return "You failed to upload " + i + " because the file was empty.";
            }
        }
        return "upload successful";
    }
}
```

8.1.3　创建文件 file.html

在 resources/templates 目录下创建文件 file.html，代码如例 8-3 所示。

【例 8-3】 创建文件 file.html 的代码示例。

```
<!DOCTYPE html>
<html xmlns="http://www.w3.org/1999/xhtml" xmlns:th="http://www.thymeleaf.org"
      xmlns:sec="http://www.thymeleaf.org/thymeleaf-extras-springsecurity3">
```

```
<head>
    <meta charset="UTF-8"/>
    <title>文件</title>
</head>
<body>
<form method="POST" enctype="multipart/form-data" action="/upload">
    <p>文件：<input type="file" name="file" /></p>
    <p><input type="submit" value="上传" /></p>
</form>
</body>
</html>
```

8.1.4 创建文件 multifile.html

在 resources/templates 目录下创建文件 multifile.html，代码如例 8-4 所示。

【例 8-4】 创建文件 multifile.html 的代码示例。

```
<!DOCTYPE html>
<html xmlns="http://www.w3.org/1999/xhtml" xmlns:th="http://www.thymeleaf.org"
      xmlns:sec="http://www.thymeleaf.org/thymeleaf-extras-springsecurity3">
<head>
    <meta charset="UTF-8"/>
    <title>多个文件上传</title>
</head>
<body>
<form method="POST" enctype="multipart/form-data" action="/batch/upload">
    <p>文件 1：<input type="file" name="file" /></p>
    <p>文件 2：<input type="file" name="file" /></p>
    <p>文件 3：<input type="file" name="file" /></p>
    <p><input type="submit" value="上传" /></p>
</form>
</body>
</html>
```

8.1.5 运行程序

运行程序后，在浏览器中输入 localhost:8080/file，结果如图 8-1 所示；选择要上传的文件，结果如图 8-2 所示；上传文件成功之后的结果，结果如图 8-3 所示。在浏览器中输入 localhost:8080/multifile，结果如图 8-4 所示；选择要上传的多个文件，结果如图 8-5 所示；多个文件上传成功后，结果如图 8-6 所示。上传文件成功后，就可以在项目的根目录下看到所上传的文件了。

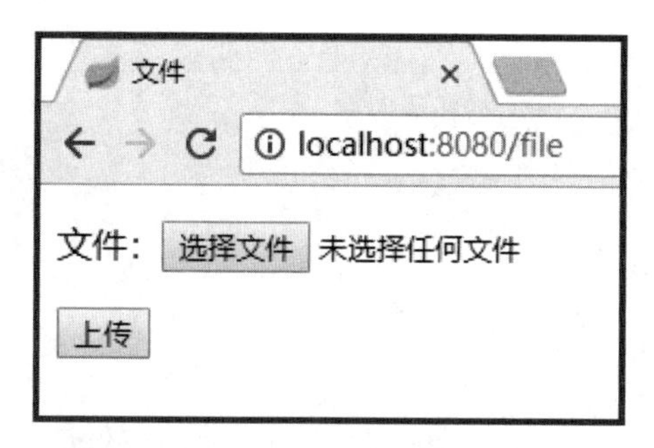

图 8-1　在浏览器中输入 localhost:8080/file 后的结果

图 8-2　选择要上传的文件后的结果

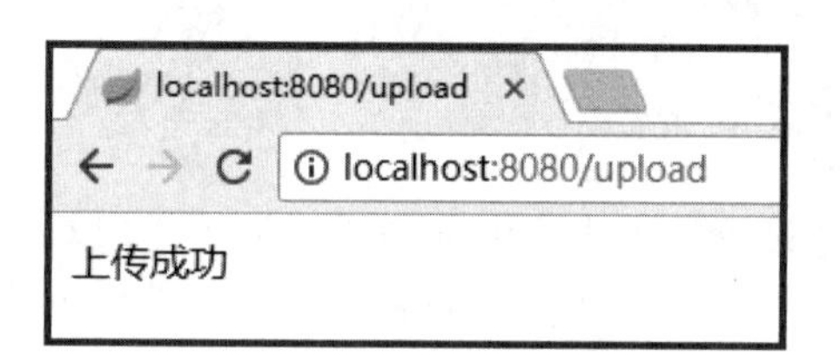

图 8-3　文件上传成功后的结果

图 8-4　在浏览器中输入 localhost:8080/multifile 后的结果

图 8-5　选择要上传的多个文件后的结果

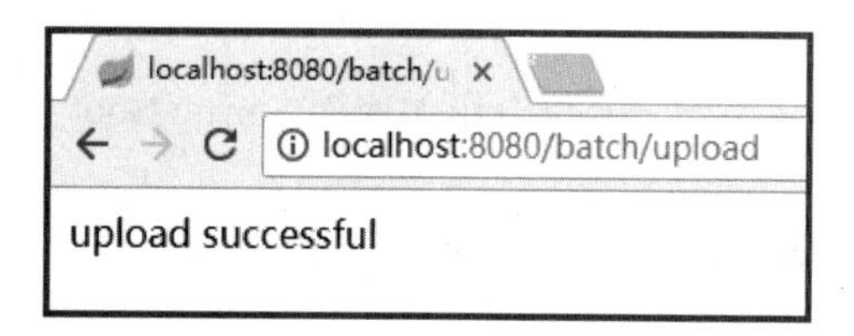

图 8-6　多个文件上传成功后的结果

8.1.6　扩展程序

修改入口类，修改后的代码如例 8-5 所示。

【例 8-5】　修改入口类的代码示例。

```
package com.bookcode;
import org.springframework.boot.SpringApplication;
import org.springframework.boot.autoconfigure.SpringBootApplication;
```

```
import org.springframework.boot.web.servlet.MultipartConfigFactory;
import org.springframework.context.annotation.Bean;
import javax.servlet.MultipartConfigElement;
@SpringBootApplication
public class DemoApplication {
  @Bean
   public MultipartConfigElement multipartConfigElement() {
       MultipartConfigFactory factory = new MultipartConfigFactory();
       //设置文件大小限制，超过设置的大小，则页面会抛出异常信息，这时候就需要进行异常信息的处理
       factory.setMaxFileSize("128KB");
       //设置上传数据的总大小
       factory.setMaxRequestSize("256KB");
       return factory.createMultipartConfig();
   }
   public static void main(String[] args) {
       SpringApplication.run(DemoApplication.class, args);
   }
}
```

运行修改后的入口类，如果上传的文件超过设置的大小，就会报错。为了实现对上传文件大小的限制，也可以不修改入口类，而在配置文件 application.properties 中设置上传文件的参数，代码如例 8-6 所示。

【例 8-6】 配置文件 application.properties 的代码示例。

```
spring.servlet.multipart.max-file-size=128KB
spring.servlet.multipart.max-request-size=256KB
```

8.2 文件下载

视频讲解

本节结合实例介绍如何用 Spring Boot 实现文件的下载。

8.2.1 添加依赖

在 pom.xml 文件中<dependencies>和</dependencies>之间添加依赖，代码如例 8-1 所示。

8.2.2 创建类 FileDownloadController

在包 com.bookcode.controller 中创建类 FileDownloadController，代码如例 8-7 所示。

【例 8-7】 创建类 FileDownloadController 的代码示例。

```
package com.bookcode.controller;
import org.springframework.stereotype.Controller;
import org.springframework.web.bind.annotation.RequestMapping;
import org.springframework.web.bind.annotation.RequestMethod;
```

```
import javax.servlet.http.HttpServletResponse;
import java.io.*;
@Controller
public class FileDownloadController {
    @RequestMapping("/downloadfile")
    public String file() {
        return "/downloadfile";    }
    @RequestMapping(value="/testDownload",method=RequestMethod.GET)
    public void testDownload(HttpServletResponse resp) throws IOException {
     String fileName = "test.txt";
     String pathName ="D:/";
        File file = new File(pathName+fileName);
        resp.setHeader("content-type", "application/octet-stream");
         resp.setContentType("application/octet-stream");
         resp.setHeader("Content-Disposition","attachment;filename="+fileName);
         byte[] buff = new byte[1024];
         BufferedInputStream bis = null;
         OutputStream os = null;
         try {
             os = resp.getOutputStream();
             bis = new BufferedInputStream(new FileInputStream(file));
             int i = bis.read(buff);
             while(i != -1) {
                 os.write(buff, 0, buff.length);
                 os.flush();
                 i = bis.read(buff);
             }
         }
         catch(IOException e) {
             e.printStackTrace();
         }
         finally {
             if(bis != null) {
                 try {
                     bis.close();
                 }
                 catch(IOException e) {
                     e.printStackTrace();
                 }
             }
        }
    }
}
```

8.2.3 创建文件 downloadfile.html

在 resources/templates 目录下创建文件 downloadfile.html，代码如例 8-8 所示。

【例 8-8】 创建文件 downloadfile.html 的代码示例。

```
<!DOCTYPE html>
<html xmlns="http://www.w3.org/1999/xhtml" xmlns:th="http://www.thymeleaf.org"
      xmlns:sec="http://www.thymeleaf.org/thymeleaf-extras-springsecurity3">
<head>
    <meta charset="UTF-8"/>
    <title>下载文件</title>
</head>
<body>
<a href="/testDownload"><h2>下载文件</h2></a>
</body>
</html>
```

8.2.4 运行程序

运行程序，在浏览器中输入 localhost:8080/downloadfile，结果如图 8-7 所示；单击“下载文件”链接，就可以下载文件（D:\test.txt）了。

图 8-7 在浏览器中输入 localhost:8080/downloadfile 后的结果

8.3 图片文件上传和显示

视频讲解

本节结合实例介绍如何用 Spring Boot 实现图片文件的上传和显示。

8.3.1 添加依赖

首先，在 pom.xml 文件中<dependencies>和</dependencies>之间添加依赖，代码如例 8-9 所示。

【例 8-9】 添加依赖的代码示例。

```
<dependency>
            <groupId>org.springframework.boot</groupId>
```

```
            <artifactId>spring-boot-starter-web</artifactId>
</dependency>
<dependency>
            <groupId>org.springframework.boot</groupId>
            <artifactId>spring-boot-starter-thymeleaf</artifactId>
</dependency>
 <dependency>
            <groupId>mysql</groupId>
            <artifactId>mysql-connector-java</artifactId>
</dependency>
<dependency>
            <groupId>org.springframework.boot</groupId>
            <artifactId>spring-boot-starter-data-jpa</artifactId>
</dependency>
```

8.3.2　创建类 User

在包 com.bookcode.entity 中创建类 User，代码如例 8-10 所示。

【例 8-10】 创建类 User 的代码示例。

```
package com.bookcode.entity;
import javax.persistence.Entity;
import javax.persistence.GeneratedValue;
import javax.persistence.Id;
import javax.persistence.Table;
@Entity
@Table(name = "t_user")
public class User {
    @Id
    @GeneratedValue
    private Long id;
    private String username;
    private String password;
    private String tupian;              //图片保存的绝对地址
    public User(){}
    public Long getId() {
        return id;
    }
    public void setId(Long id) {
        this.id = id;
    }
    public String getUsername() {
        return username;
    }
    public void setUsername(String username) {
```

```
        this.username = username;
    }
    public String getPassword() {
        return password;
    }
    public void setPassword(String password) {
        this.password = password;
    }
    public String getTupian() {
        return tupian;
    }
    public void setTupian(String tupian) {
        this.tupian = tupian;
    }
}
```

8.3.3 创建接口 UserRepository

在包 com.bookcode.dao 中创建接口 UserRepository，代码如例 8-11 所示。

【例 8-11】 创建接口 UserRepository 的代码示例。

```
package com.bookcode.dao;
import com.bookcode.entity.User;
import org.springframework.data.jpa.repository.JpaRepository;
public interface UserRepository extends JpaRepository<User,Long> {
}
```

8.3.4 创建类 MyWebConfig

在包 com.bookcode.config 中创建类 MyWebConfig，代码如例 8-12 所示。

【例 8-12】 创建类 MyWebConfig 的代码示例。

```
package com.bookcode.config;
import org.springframework.context.annotation.Configuration;
import org.springframework.web.servlet.config.annotation.Resource
HandlerRegistry;
import org.springframework.web.servlet.config.annotation.
WebMvcConfigurerAdapter;
@Configuration
public class MyWebConfig extends WebMvcConfigurerAdapter {
    @Override
    public void addResourceHandlers(ResourceHandlerRegistry registry) {
    registry.addResourceHandler("/src/main/webapp/**").addResourceLocations
    ("classpath:/webapp/");
```

```
        super.addResourceHandlers(registry);
    }
}
```

8.3.5 创建类 UserPictureController

在包 com.bookcode.controller 中创建类 UserPictureController，代码如例 8-13 所示。

【例 8-13】 创建类 UserPictureController 的代码示例。

```
package com.bookcode.controller;
import com.bookcode.dao.UserRepository;
import com.bookcode.entity.User;
import org.springframework.beans.factory.annotation.Autowired;
import org.springframework.stereotype.Controller;
import org.springframework.ui.Model;
import org.springframework.web.bind.annotation.GetMapping;
import org.springframework.web.bind.annotation.PostMapping;
import org.springframework.web.bind.annotation.RequestParam;
import org.springframework.web.multipart.MultipartFile;
import java.io.*;
@Controller
public class UserPictureController {
    @Autowired
    UserRepository userRepository;
    @GetMapping(value="/zhuce")
    public String zhuce(){
        return "zhuce";
    }
    @PostMapping(value="/zhuce")
    public String tijiao(@RequestParam(value="name") String name,
                         @RequestParam(value="password") String password,
                         @RequestParam(value="file")MultipartFile file,
                         Model model) {
        User user = new User();
        user.setUsername(name);
        user.setPassword(password);
        if(!file.isEmpty()) {
          try {
  BufferedOutputStream out = new BufferedOutputStream( new FileOutputStream(new
              File("D:\\ch83\\src\\main\\webapp\\"+name+".jpg")));
                                                  //保存图片到目录下
              out.write(file.getBytes());
              out.flush();
              out.close();
              String filename="D:\\ch83\\src\\main\\webapp\\"+name+".jpg";
              user.setTupian(filename);
```

```
                userRepository.save(user);         //增加用户
            }
            catch(FileNotFoundException e) {
                e.printStackTrace();
                return"上传失败," + e.getMessage();
            }
            catch(IOException e) {
                e.printStackTrace();
                return "上传失败," + e.getMessage();
            }
            model.addAttribute(user);
            return "permanager";
        }
        else {
            return "上传失败，因为文件是空的.";
        }
    }
}
```

8.3.6 创建文件 zhuce.html

创建文件 zhuce.html，代码如例 8-14 所示。

【例 8-14】 创建文件 zhuce.html 的代码示例。

```
<!DOCTYPE html>
<html lang="en" xmlns:th="http://www.thymeleaf.org">
<head>
    <meta charset="UTF-8"/>
    <title>注册页面</title>
</head>
<body>
<form  action="/zhuce"  th:action="@{/zhuce}"  method="post"  enctype=
"multipart/form-data" >
    <label>姓名</label><input type="text" name="name"/>
    <label>密码</label><input type="password" name="password"/>
    <label>上传图片</label>
    <input type="file" name="file"/>
    <input type="submit" value="上传"/>
</form>
</body>
</html>
```

8.3.7 创建文件 permanager.html

创建文件 permanager.html，代码如例 8-15 所示。

【例 8-15】 创建文件 permanager.html 的代码示例。

```
<!DOCTYPE html>
<html lang="en" xmlns:th="http://www.thymeleaf.org">
<head>
    <meta charset="UTF-8"/>
    <title>个人中心</title>
</head>
<body>
<p>用户名:</p>
<p th:text="${user.username}"></p>
<p>图片:</p>
<img th:src="@{${user.username}+'.jpg'}"/>
</body>
</html>
```

8.3.8　创建配置文件 application.yml

创建配置文件 application.yml，代码如例 8-16 所示。

【例 8-16】 创建配置文件 application.yml 的代码示例。

```
spring:
    datasource:
      driver-class-name: com.mysql.jdbc.Driver
      url: jdbc:mysql://localhost:3306/mytest
      username: root
      password: sa
    jpa:
      hibernate:
        ddl-auto: update
      show-sql: true
```

8.3.9　创建目录并运行程序

在目录 src/main 下创建子目录 webapp。

运行程序，在浏览器中输入 localhost:8080/zhuce 并选择要上传的文件，结果如图 8-8 所示。单击“上传”按钮，成功上传图片文件并显示文件后的结果如图 8-9 所示。

图 8-8　在浏览器中输入 localhost:8080/zhuce
并选择要上传的文件后的结果

图 8-9 成功上传图片文件并显示文件后的结果

8.4 访问 HDFS

视频讲解

Hadoop 分布式文件系统（Hadoop Distributed File System，HDFS）被设计成适合运行在通用硬件上的分布式文件系统。HDFS 是一个高度容错性的系统，适合部署在廉价的机器上。HDFS 能提供高吞吐量的数据访问，非常适合大规模数据集上的应用。HDFS 是 Apache Hadoop 项目的一部分。HDFS 提供高吞吐量来访问应用程序的数据，适合那些有着大数据集的应用程序。本节结合实例介绍如何使用 Spring Boot 访问 HDFS。

8.4.1 添加依赖

首先，在 pom.xml 文件中<dependencies>和</dependencies>之间添加依赖，代码如例 8-17 所示。

【例 8-17】 添加依赖的代码示例。

```
<dependency>
        <groupId>org.springframework.data</groupId>
        <artifactId>spring-data-hadoop-boot</artifactId>
        <version>2.5.0.RELEASE</version>
</dependency>
<dependency>
        <groupId>org.apache.hadoop</groupId>
        <artifactId>hadoop-client</artifactId>
        <version>2.5.2</version>
        <scope>provided</scope>
</dependency>
```

8.4.2　修改入口类

修改入口类 DemoApplication，代码如例 8-18 所示。

【例 8-18】 入口类 DemoApplication 的代码示例。

```
package com.bookcode;
import org.springframework.beans.factory.annotation.Autowired;
import org.springframework.boot.CommandLineRunner;
import org.springframework.boot.SpringApplication;
import org.springframework.boot.autoconfigure.SpringBootApplication;
import org.apache.hadoop.fs.FileStatus;
import org.springframework.data.hadoop.fs.FsShell;
@SpringBootApplication
public class DemoApplication implements CommandLineRunner {
    @Autowired
    FsShell fsShell;  //用于执行 HDFS shell 命令的对象
    public void run(String... strings) throws Exception {
        //查看根目录下的所有文件
        for(FileStatus fileStatus : fsShell.ls("/")) {
            System.out.println("> " + fileStatus.getPath());
        }
    }
    public static void main(String[] args) {
        SpringApplication.run(DemoApplication.class, args);
    }
}
```

8.4.3　运行程序

运行程序，在控制台上输出 d:\下的所有子目录，如图 8-10 所示。

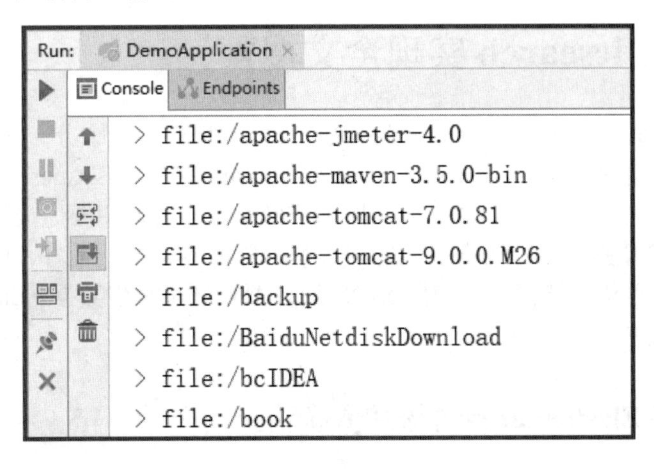

图 8-10　控制台上输出的 d:\下的所有子目录信息

8.4.4 简化程序

去掉在 pom.xml 文件中所添加的如例 8-17 所示的依赖。修改入口类 DemoApplication，代码如例 8-19 所示。

【例 8-19】 修改入口类 DemoApplication 的代码示例。

```
package com.bookcode;
import org.springframework.boot.CommandLineRunner;
import org.springframework.boot.SpringApplication;
import org.springframework.boot.autoconfigure.SpringBootApplication;
import org.springframework.util.ResourceUtils;
import java.io.File;
@SpringBootApplication
public class DemoApplication implements CommandLineRunner {
        public void run(String... strings) throws Exception {
        File file = ResourceUtils.getFile("file:d:\\");
                if(file.exists()){
                    File[] files = file.listFiles();
                    if(files != null){
                        for(File childFile:files){
                            System.out.println(">file:/"+childFile.getName());
                        }
                    }
                }
        }
        public static void main(String[] args) {
                SpringApplication.run(DemoApplication.class, args);
        }
}
```

运行程序，在控制台上输出 d:\下的所有子目录，如图 8-10 所示。

8.5 用 Elasticsearch 实现全文搜索

视频讲解

Elasticsearch 是一个基于 Lucene 的搜索服务器。它提供了一个分布式多用户功能的全文搜索引擎和基于 RESTful 风格的 Web 接口。Elasticsearch 是当前流行的企业级搜索引擎。建立一个网站或应用程序时，想要完成搜索工作的创建比较困难。开发者希望搜索解决方案要运行速度快、零配置、扩展性好、服务可靠性高、实时性好、免费，还希望能够简单地使用 JSON 通过 HTTP 来索引数据。Elasticsearch 可以较好地帮助开发者完成这些目标。

8.5.1 安装 Elasticsearch 并添加依赖

首先，下载并正确安装 Elasticsearch，再在 pom.xml 文件中<dependencies>和</dependencies>

之间添加依赖，代码如例 8-20 所示。

【例 8-20】 添加依赖的代码示例。

```
<dependency>
          <groupId>org.springframework.boot</groupId>
          <artifactId>spring-boot-starter-data-elasticsearch</artifactId>
</dependency>
```

8.5.2　创建类 EsBlog

在包 com.bookcode.entity 中创建类 EsBlog，代码如例 8-21 所示。

【例 8-21】 创建类 EsBlog 的代码示例。

```
package com.bookcode.entity;
import java.io.Serializable;
import org.springframework.data.annotation.Id;
import org.springframework.data.elasticsearch.annotations.Document;
@Document(indexName = "blog", type = "blog") //文档
public class EsBlog implements Serializable {
    @Id //主键
    private String id;
    private String title;
    private String summary;
    private String content;
    protected EsBlog() { //JPA 的规范要求无参构造函数，设为 protected 以防止直接使用
    }
    public EsBlog(String title, String summary, String content) {
        this.title = title;
        this.summary = summary;
        this.content = content;
    }
    public String getId() {
        return id;
    }
    public void setId(String id) {
        this.id = id;
    }
    public String getTitle() {
        return title;   }
    public void setTitle(String title) {
        this.title = title;
    }
    public String getSummary() {
        return summary;
    }
```

```
    public void setSummary(String summary) {
        this.summary = summary;
    }
    public String getContent() {
        return content;
    }
    public void setContent(String content) {
        this.content = content;
    }
    @Override
    public String toString() {
        return String.format(
                "EsBlog[id='%s',title='%s',summary='%s',content='%s']",
                id, title, summary, content);
    }
}
```

8.5.3 创建接口 EsBlogRepository

在包 com.bookcode.dao 中创建接口 EsBlogRepository，代码如例 8-22 所示。

【例 8-22】 创建接口 EsBlogRepository 的代码示例。

```
package com.bookcode.dao;
import com.bookcode.entity.EsBlog;
import org.springframework.data.domain.Page;
import org.springframework.data.domain.Pageable;
import org.springframework.data.elasticsearch.repository.ElasticsearchRepository;
public interface EsBlogRepository extends ElasticsearchRepository<EsBlog,String> {
    //分页查询博客
Page<EsBlog> findByTitleContainingOrSummaryContainingOrContentContaining
(String title, String summary, String content, Pageable pageable);
}
```

8.5.4 创建类 EsBlogRepositoryTest

创建类 EsBlogRepositoryTest，代码如例 8-23 所示。

【例 8-23】 创建类 EsBlogRepositoryTest 的代码示例。

```
package com.bookcode;
import com.bookcode.dao.EsBlogRepository;
import com.bookcode.entity.EsBlog;
import org.junit.Before;
import org.junit.Test;
import org.junit.runner.RunWith;
import org.springframework.beans.factory.annotation.Autowired;
```

```
import org.springframework.boot.test.context.SpringBootTest;
import org.springframework.data.domain.Page;
import org.springframework.data.domain.PageRequest;
import org.springframework.data.domain.Pageable;
import org.springframework.test.context.junit4.SpringRunner;
@RunWith(SpringRunner.class)
@SpringBootTest
public class EsBlogRepositoryTest {
    @Autowired
    private EsBlogRepository esBlogRepository;
    @Before
    public void initRepositoryData() {
        esBlogRepository.deleteAll();  //清除所有数据
        //初始化数据
        esBlogRepository.save(new EsBlog("Had I not seen the Sun",
            "I could have borne the shade",
              "But Light a newer Wilderness. My Wilderness has made."));
       esBlogRepository.save(new EsBlog("There is room in the halls of pleasure",
            "For a long and lordly train",
              "But one by one we must all file on, Through the narrow aisles
              of pain."));
       esBlogRepository.save(new EsBlog("When you are old",
            "When you are old and grey and full of sleep",
              "And nodding by the fire, take down this book."));
    }
    @Test
public void testFindDistinctEsBlogByTitleContainingOrSummaryContaining-
OrContentContaining() {
        Pageable pageable = PageRequest.of(0, 20);
        String title = "Sun";
        String summary = "is";
        String content = "down";
Page<EsBlog> page = esBlogRepository.findByTitleContainingOrSummary-
ContainingOrContentContaining
(title, summary, content, pageable);
        System.out.println("---------start 1");
        for(EsBlog blog : page) {
            System.out.println(blog.toString());
        }
        System.out.println("---------end 1");
        title = "the";
        summary = "the";
        content = "the";
    page = esBlogRepository.findByTitleContainingOrSummary-
    ContainingOrContentContaining
```

```
(title, summary, content, pageable);
        System.out.println("---------start 2");
        for(EsBlog blog : page) {
            System.out.println(blog.toString());
        }
        System.out.println("---------end 2");
    }
}
```

8.5.5 修改配置文件 application.properties

修改配置文件 application.properties，代码如例 8-24 所示。

【例 8-24】 修改配置文件 application.properties 的代码示例。

```
#Elasticsearch 服务地址
spring.data.elasticsearch.cluster-nodes=localhost:9300
#设置连接超时时间
spring.data.elasticsearch.properties.transport.tcp.connect_timeout = 120s
```

8.5.6 运行程序（1）

为了启动 Elasticsearch，到安装目录下运行 bin\elasticsearch.bat。

运行程序中该测试类，控制台中的输出结果如图 8-11 所示。输出结果显示，第一次以 Sun、is 和 down 作为查询参数，一共有两条匹配数据（第一首诗歌中含有 Sun、第三首诗歌中含有 down）。第二次以 the 作为查询参数，一共有三条匹配数据（三首诗歌中都含有 the）。

```
---------start 1
EsBlog[id='oC4D-GMBkM19PgbvToAn',title='When you are old',summary='When you are old and grey and full of sleep',content='And no
EsBlog[id='ni4D-GMBkM19PgbvTIBD',title='Had I not seen the Sun',summary='I could have borne the shade',content='But Light a new
---------end 1
---------start 2
EsBlog[id='ny4D-GMBkM19PgbvTYCq',title='There is room in the halls of pleasure',summary='For a long and lordly train',content='
EsBlog[id='ni4D-GMBkM19PgbvTIBD',title='Had I not seen the Sun',summary='I could have borne the shade',content='But Light a new
EsBlog[id='oC4D-GMBkM19PgbvToAn',title='When you are old',summary='When you are old and grey and full of sleep',content='And no
---------end 2
```

图 8-11 控制台中的输出结果

8.5.7 创建类 BlogController

在包 com.bookcode.controller 中创建类 BlogController，代码如例 8-25 所示。

【例 8-25】 创建类 BlogController 的代码示例。

```
package com.bookcode.controller;
import java.util.List;
```

```
import com.bookcode.dao.EsBlogRepository;
import com.bookcode.entity.EsBlog;
import org.springframework.beans.factory.annotation.Autowired;
import org.springframework.data.domain.Page;
import org.springframework.data.domain.PageRequest;
import org.springframework.data.domain.Pageable;
import org.springframework.web.bind.annotation.GetMapping;
import org.springframework.web.bind.annotation.RequestMapping;
import org.springframework.web.bind.annotation.RequestParam;
import org.springframework.web.bind.annotation.RestController;
@RestController
@RequestMapping("/blogs")
public class BlogController {
    @Autowired
    private EsBlogRepository esBlogRepository;
    @GetMapping
public List<EsBlog> list(@RequestParam(value="title",required=false,
defaultValue="") String title,
            @RequestParam(value="summary",required=false,defaultValue="")
            String summary,@RequestParam(value="content",required=false,
            defaultValue="") String content, @RequestParam(value=
            "pageIndex",required=false,defaultValue="0") int pageIndex,
            @RequestParam(value="pageSize",required=false,
            defaultValue="10") int pageSize) {
        //数据在 Test 里面先初始化了，这里只管取数据
        Pageable pageable = PageRequest.of(pageIndex, pageSize);
Page<EsBlog> page = esBlogRepository.findByTitleContainingOrSummary-
ContainingOrContentContaining
(title, summary, content, pageable);
        return page.getContent();
    }
}
```

8.5.8　运行程序（2）

运行程序，为了查询 id 中包含 i、summary 中包含 love 和 content 中包含 you 的博客信息，在浏览器中输入 localhost:8080/blogs?id=i&summary=love&content=you，输出结果如图 8-12 所示。

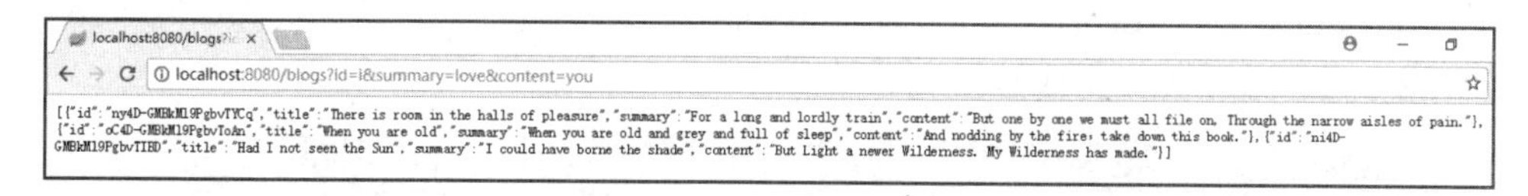

图 8-12　在浏览器中输入 localhost:8080/blogs?id=i&summary=love&content=you 后的结果

8.6 实现邮件发送

视频讲解

现在大部分网站在用户注册时都要求用户填写邮箱进行注册验证，因此发送邮件是网站必备的功能之一。在没有框架之前，一般是通过 Java 自带的 JavaMail 类来发送邮件，后来 Spring 推出了 JavaMailSender 类大大简化了发送邮件的过程，现在的 Spring Boot 又将其封装成了 spring-boot-starter-mail 模块。

8.6.1 登录邮箱并开启授权码

以 163 邮箱为例，按照图 8-13～图 8-15 依次开启邮箱授权码。

图 8-13　选择“设置”，进行 POP3 等设置

图 8-14　单击“客户端授权密码”

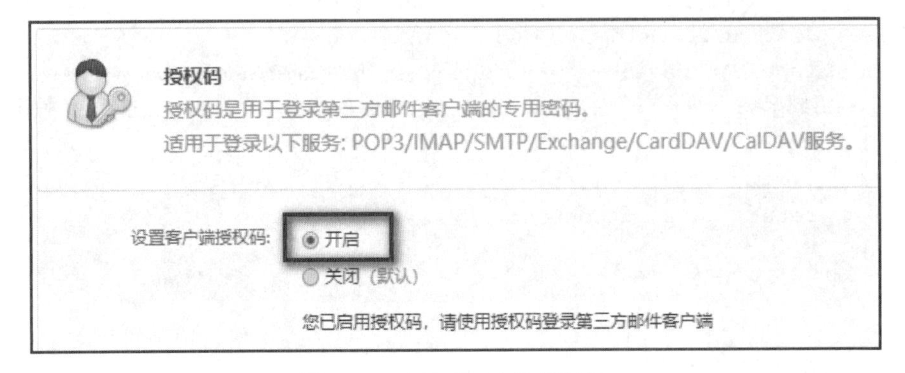

图 8-15　选择“开启”授权码

8.6.2 添加依赖

在 pom.xml 文件中<dependencies>和</dependencies>之间添加依赖，代码如例 8-26 所示。

【例 8-26】 添加依赖的代码示例。

```
<dependency>
        <groupId>org.springframework.boot</groupId>
        <artifactId>spring-boot-starter-mail</artifactId>
```

```
</dependency>
<dependency>
        <groupId>org.springframework</groupId>
        <artifactId>spring-web</artifactId>
        <version>5.0.9.RELEASE</version>
</dependency>
<dependency>
        <groupId>junit</groupId>
        <artifactId>junit</artifactId>
        <version>4.12</version>
        <scope>test</scope>
</dependency>
<dependency>
        <groupId>org.springframework</groupId>
        <artifactId>spring-test</artifactId>
        <version>5.0.6.RELEASE</version>
        <scope>test</scope>
</dependency>
<dependency>
        <groupId>org.springframework.boot</groupId>
        <artifactId>spring-boot-test</artifactId>
        <version>2.0.2.RELEASE</version>
        <scope>test</scope>
</dependency>
<dependency>
        <groupId>org.springframework.boot</groupId>
        <artifactId>spring-boot-starter-thymeleaf</artifactId>
</dependency>
```

8.6.3　创建接口 EmailService

创建接口 EmailService，代码如例 8-27 所示。

【例 8-27】 创建接口 EmailService 的代码示例。

```
package com.bookcode.service;
public interface EmailService {
   //发送简单邮件
   public void sendSimpleEmail(String to, String subject, String content);
   //发送 HTML 格式邮件
   public void sendHtmlEmail(String to, String subject, String content);
   //发送带附件的邮件
   public void sendAttachmentsEmail(String to, String subject, String
   content, String filePath);
   //发送带静态资源的邮件
   public void sendInlineResourceEmail(String to, String subject, String
```

```
    content, String rscPath, String rscId);
}
```

8.6.4 创建类 EmailServiceImp

创建类 EmailServiceImp，代码如例 8-28 所示。

【例 8-28】 创建类 EmailServiceImp 的代码示例。

```
package com.bookcode.service.impl;
import org.slf4j.Logger;
import org.slf4j.LoggerFactory;
import org.springframework.beans.factory.annotation.Autowired;
import org.springframework.beans.factory.annotation.Value;
import org.springframework.core.io.FileSystemResource;
import org.springframework.mail.SimpleMailMessage;
import org.springframework.mail.javamail.JavaMailSender;
import org.springframework.mail.javamail.MimeMessageHelper;
import org.springframework.stereotype.Component;
import javax.mail.MessagingException;
import javax.mail.internet.MimeMessage;
import java.io.File;
@Component
public class EmailServiceImp implements EmailService {
    private final Logger logger = LoggerFactory.getLogger(this.getClass());
    @Autowired
    private JavaMailSender mailSender;    //Spring 提供的邮件发送类
    @Value("${spring.mail.username}")
    private String from;
    @Override
    public void sendSimpleEmail(String to, String subject, String content) {
        SimpleMailMessage message = new SimpleMailMessage(); //创建简单邮件消息
        message.setFrom(from);                               //设置发送人
        message.setTo(to);                                   //设置收件人
        message.setSubject(subject);                         //设置主题
        message.setText(content);                            //设置内容
        try {
            mailSender.send(message);                        //执行发送邮件
            logger.info("简单邮件已经发送。");
        }
        catch(Exception e) {
            logger.error("发送简单邮件时发生异常！", e);
        }
    }
    @Override
    public void sendHtmlEmail(String to, String subject, String content) {
```

```
    MimeMessage message = mailSender.createMimeMessage();  //创建一个MINE消息
    try {
        //true表示需要创建一个multipart message
        MimeMessageHelper helper = new MimeMessageHelper(message, true);
        helper.setFrom(from);
        helper.setTo(to);
        helper.setSubject(subject);
        helper.setText(content, true);
        mailSender.send(message);
        logger.info("HTML邮件发送成功");
    }
    catch(MessagingException e) {
        logger.error("发送HTML邮件时发生异常！", e);
    }
}
@Override
public void sendAttachmentsEmail(String to, String subject, String
content, String filePath) {
    MimeMessage message = mailSender.createMimeMessage();  //创建一个MINE消息
    try {
        MimeMessageHelper helper = new MimeMessageHelper(message, true);
        helper.setFrom(from);
        helper.setTo(to);
        helper.setSubject(subject);
        helper.setText(content, true);    //true表示这个邮件是有附件的
    FileSystemResource file = new FileSystemResource(new File(filePath));
                                          //创建文件系统资源
        String fileName = filePath.substring(filePath.lastIndexOf (File.
        separator));
        helper.addAttachment(fileName, file);  //添加附件
        mailSender.send(message);
        logger.info("带附件的邮件已经发送。");
    }
    catch(MessagingException e) {
        logger.error("发送带附件的邮件时发生异常！", e);
    }
}
@Override
public void sendInlineResourceEmail(String to, String subject, String
content, String rscPath, String rscId) {
    MimeMessage message = mailSender.createMimeMessage();
    try {
        MimeMessageHelper helper = new MimeMessageHelper(message, true);
        helper.setFrom(from);
        helper.setTo(to);
```

```
            helper.setSubject(subject);
            helper.setText(content, true);
            FileSystemResource res = new FileSystemResource(new File(rscPath));
            //添加内联资源，一个 id 对应一个资源，最终通过 id 来找到该资源
            helper.addInline(rscId, res);
            //添加多个图片可以使用多条 <img src='cid:" + rscId + "' > 和 helper.
            //addInline(rscId, res) 来实现
            mailSender.send(message);
            logger.info("嵌入静态资源的邮件已经发送。");
        }
        catch(MessagingException e) {
            logger.error("发送嵌入静态资源的邮件时发生异常！", e);
        }
    }
}
```

8.6.5 创建类 DemoApplicationTests

创建类 DemoApplicationTests，代码如例 8-29 所示。

【例 8-29】 创建类 DemoApplicationTests 的代码示例。

```
package com.bookcode;
import org.junit.Test;
import org.junit.runner.RunWith;
import org.springframework.beans.factory.annotation.Autowired;
import org.springframework.boot.test.context.SpringBootTest;
import org.springframework.test.context.junit4.SpringRunner;
import org.thymeleaf.TemplateEngine;
import org.thymeleaf.context.Context;
@RunWith(SpringRunner.class)
@SpringBootTest
public class DemoApplicationTests {
    @Test
    public void contextLoads() {
    }
    @Autowired
    private EmailService emailService;
    @Test
    public void sendSimpleMail() throws Exception {
        emailService.sendSimpleEmail("97291@qq.com","this is simple
        mail"," hello 张三");
    }
    @Test
    public void sendHtmlMail() throws Exception {
        String content="<html>\n" +
```

```
                "<body>\n" +
                "    <h3>hello world ! 这是一封 HTML 邮件!</h3>\n" +
                "</body>\n" +
                "</html>";
        emailService.sendHtmlEmail("97291@qq.com","this is html mail",content);
    }
    @Test
    public void sendAttachmentsMail() {
        String filePath="d:\\pk.png";
        emailService.sendAttachmentsEmail("97291@qq.com", "主题：带附件的邮件",
        "收到附件，请查收！", filePath);
    }
    @Test
    public void sendInlineResourceMail() {
        String rscId = "001";
        String content="<html><body>这是有图片的邮件：<img src=\'cid:" + rscId
        + "\' ></body></html>";
        String imgPath = "d:\\pk.png";
        emailService.sendInlineResourceEmail("97291@qq.com", "主题：这是有图
        片的邮件", content, imgPath, rscId);
    }
    @Autowired
    private TemplateEngine templateEngine;
    @Test
    public void sendTemplateMail() {
        //创建邮件正文
        Context context = new Context();
        context.setVariable("username", "张三");
        String emailContent = templateEngine.process("email", context);
        System.out.println(emailContent);
        emailService.sendHtmlEmail("97291@qq.com","主题:这是模板邮件",emailContent);
    }
}
```

8.6.6 修改配置文件 application.properties

修改配置文件 application.properties，代码如例 8-30 所示。

【例 8-30】 修改配置文件 application.properties 的代码示例。

```
#邮箱服务器地址
spring.mail.host=smtp.163.com
#用户名
#请修改成您自己的邮箱和授权码
spring.mail.username=ws@163.com
#授权密码
```

```
spring.mail.password=w1234567
spring.mail.default-encoding=UTF-8
```

8.6.7 创建文件 email.html

创建文件 email.html，代码如例 8-31 所示。

【例 8-31】 创建文件 email.html 的代码示例。

```
<!DOCTYPE html>
<html lang="zh" xmlns:th="http://www.thymeleaf.org">
<head>
    <meta charset="UTF-8"/>
    <title>Title</title>
</head>
<body>
<h4 th:text="|尊敬的${username}:|"></h4><br /><br />
感谢您对 Spring Boot 的关注与支持。<br />
</body>
</html>
```

8.6.8 运行程序

运行程序中测试类，则在收件人的邮箱中依次收到 HTML、模板、简单、带附件、有图片的邮件 5 封，如图 8-16 所示。5 封邮件的内容依次如图 8-17～图 8-21 所示。

主题：这是有图片的邮件 - 这是有图片的邮件：
主题：带附件的邮件 - 收到附件，请查收！
this is simple mail - hello 张三
主题：这是模板邮件 - 尊敬的 张三：感谢您对Spring Boot的关注与支持。
this is html mail - hello world！这是一封HTML邮件！

图 8-16 收到 5 封邮件的信息

hello world！这是一封HTML邮件！

图 8-17 this is html mail 的邮件内容

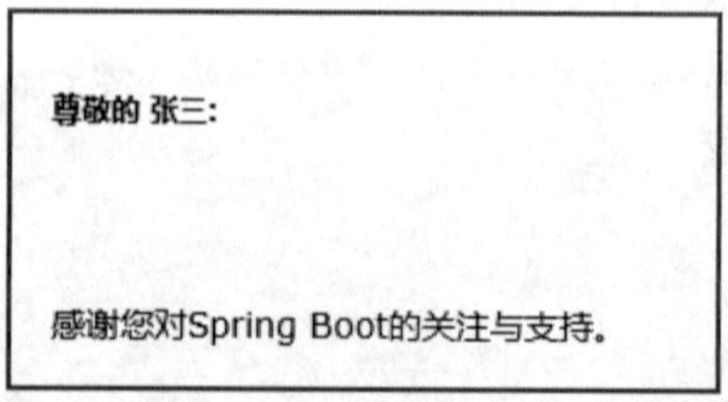

尊敬的 张三：

感谢您对Spring Boot的关注与支持。

图 8-18 主题为“这是模板邮件”的邮件内容

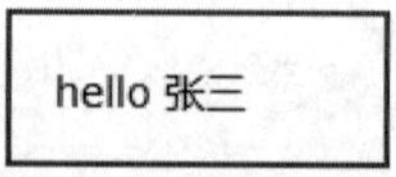

hello 张三

图 8-19 this is simple mail 的邮件内容

收到附件，请查收！

附件(1 个)

普通附件

pk.png (254.17K)

预览　下载　收藏　转存

图 8-20　主题为“带附件的邮件”的邮件内容

图 8-21　主题为“这是有图片的邮件”的邮件内容

8.7　用 REST Docs 创建 API 文档

视频讲解

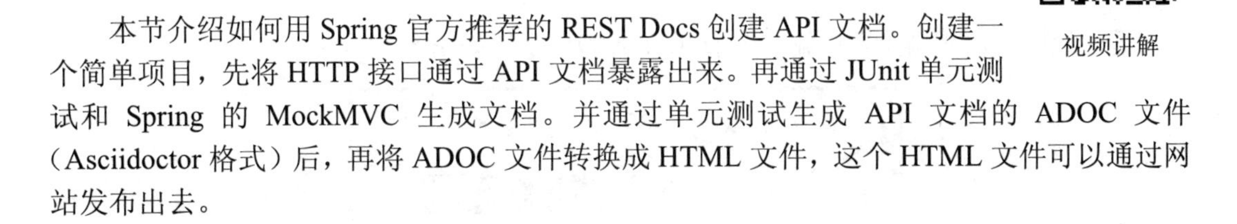

本节介绍如何用 Spring 官方推荐的 REST Docs 创建 API 文档。创建一个简单项目，先将 HTTP 接口通过 API 文档暴露出来。再通过 JUnit 单元测试和 Spring 的 MockMVC 生成文档。并通过单元测试生成 API 文档的 ADOC 文件（Asciidoctor 格式）后，再将 ADOC 文件转换成 HTML 文件，这个 HTML 文件可以通过网站发布出去。

8.7.1　添加依赖

在 pom.xml 文件中<dependencies>和</dependencies>之间添加依赖，代码如例 8-32

所示。

【例 8-32】 添加依赖的代码示例。

```
<dependency>
        <groupId>org.springframework.boot</groupId>
        <artifactId>spring-boot-starter-web</artifactId>
</dependency>
<dependency>
        <groupId>org.springframework.restdocs</groupId>
        <artifactId>spring-restdocs-mockmvc</artifactId>
        <scope>test</scope>
</dependency>
```

8.7.2 创建类 HomeController

创建类 HomeController，代码如例 8-33 所示。

【例 8-33】 创建类 HomeController 的代码示例。

```
package com.bookcode.controller;
import org.springframework.web.bind.annotation.GetMapping;
import org.springframework.web.bind.annotation.RestController;
import java.util.Collections;
import java.util.Map;
@RestController
public class HomeController {
   @GetMapping("/")
   public Map<String, Object> greeting() {
       return Collections.singletonMap("message", "Hello World");
   }
}
```

8.7.3 运行程序

运行程序后，在浏览器中输入 localhost:8080，结果如图 8-22 所示。

图 8-22 在浏览器中输入 localhost:8080 后的结果

8.7.4 创建类 WebLayerTest

创建类 WebLayerTest，代码如例 8-34 所示。

【例 8-34】 创建类 WebLayerTest 的代码示例。

```
package com.bookcode;
import com.bookcode.controller.HomeController;
import org.junit.Test;
import org.junit.runner.RunWith;
import org.springframework.beans.factory.annotation.Autowired;
import org.springframework.boot.test.autoconfigure.restdocs.
AutoConfigureRestDocs;
import org.springframework.boot.test.autoconfigure.web.servlet.WebMvcTest;
import org.springframework.test.context.junit4.SpringRunner;
import org.springframework.test.web.servlet.MockMvc;
import static org.hamcrest.Matchers.containsString;
import static org.springframework.restdocs.mockmvc.
MockMvcRestDocumentation.document;
import static org.springframework.test.web.servlet.request.
MockMvcRequestBuilders.get;
import static org.springframework.test.web.servlet.result.MockMvcResult-
Handlers.print;
import static org.springframework.test.web.servlet.result.MockMvcResult-
Matchers.content;
import static org.springframework.test.web.servlet.result.MockMvcResult-
Matchers.status;
@RunWith(SpringRunner.class)
@WebMvcTest(HomeController.class)
@AutoConfigureRestDocs(outputDir = "target/snippets")
public class WebLayerTest {
    @Autowired
    private MockMvc mockMvc;
    @Test
    public void shouldReturnDefaultMessage() throws Exception {
        this.mockMvc.perform(get("/")).andDo(print()).andExpect(status().isOk())
                .andExpect(content().string(containsString("Hello World")))
                .andDo(document("home"));
    }
}
```

启动单元测试，测试通过，发现在 target 文件夹下生成了一个 snippets 文件夹，目录下还生成了一些 ADOC 文件（Asciidoctor 格式），如图 8-23 所示。

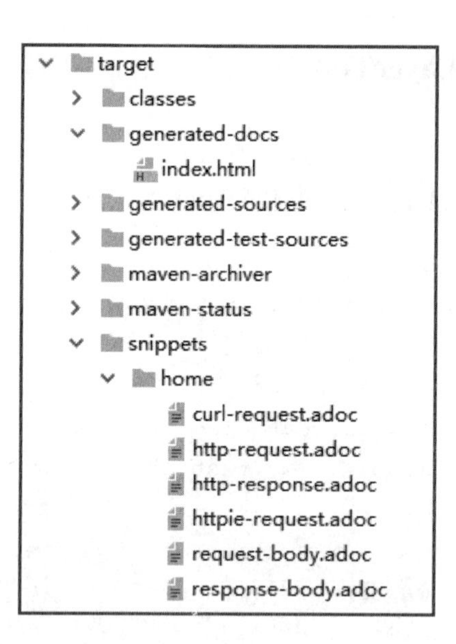

图 8-23 自动生成的 ADOC 文件

8.7.5 创建文件 index.adoc

在 src/main/asciidoc 目录下创建文件 index.adoc，代码如例 8-35 所示。

【例 8-35】 创建文件 index.adoc 的代码示例。

```
= 用 Spring REST Docs 创建文档
This is an example output for a service running at http://localhost:8080:
.request
include::{snippets}/home/http-request.adoc[]
.response
include::{snippets}/home/http-response.adoc[]
这个例子非常简单，通过单元测试和一些简单的配置就能够得到 api 文档了。
```

8.7.6 添加插件

在 pom.xml 文件中<plugins>和</plugins>之间添加插件 asciidoctor-maven-plugin，代码如例 8-36 所示。

【例 8-36】 添加插件 asciidoctor-maven-plugin 的代码示例。

```
<plugin>
            <groupId>org.asciidoctor</groupId>
            <artifactId>asciidoctor-maven-plugin</artifactId>
            <executions>
               <execution>
```

```
                        <id>generate-docs</id>
                        <phase>prepare-package</phase>
                        <goals>
                            <goal>process-asciidoc</goal>
                        </goals>
                        <configuration>
                            <sourceDocumentName>index.adoc</sourceDocumentName>
                            <backend>html</backend>
                            <attributes>
                                <snippets>${project.build.directory}/snippets
                                 </snippets>
                            </attributes>
                        </configuration>
                    </execution>
                </executions>
</plugin>
```

8.7.7　利用 Maven 的 package 命令生成文件

可以利用 Maven 的 package 命令（如图 8-24 所示）根据 index.adoc 文件生成文件 index.html。

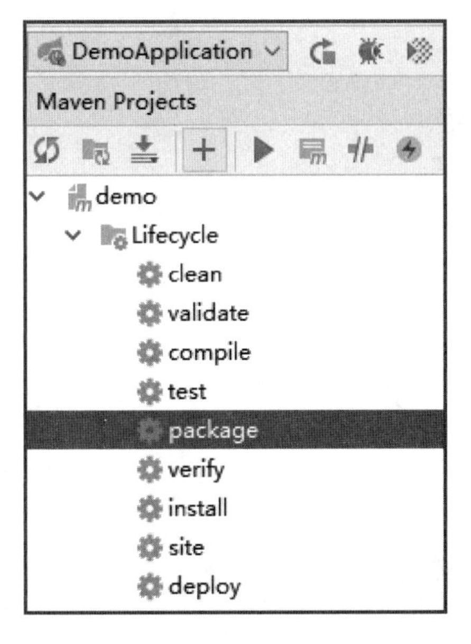

图 8-24　Maven 的 package 命令

自动生成的文件 index.html 在/target/generated-docs 目录下。用浏览器打开文件 index.html，结果如图 8-25 所示。在浏览器中单击 index.html 中的链接 http://localhost:8080，跳转到如图 8-22 所示的界面。

图 8-25　用浏览器中打开文件 index.html 的结果

习题 8

简答题

1．简述对 HDFS 的理解。

2．简述对 Elasticsearch 的理解。

实验题

1．实现文件上传。

2．实现文件下载。

3．实现图片文件上传和显示。

4．实现输出 d 盘下文件名。

5．利用 Elasticsearch 实现全文搜索。

6．实现不同类型邮件的发送。

7．实现用 REST Docs 创建 API 文档。

第 9 章

Spring Boot 的 WebFlux 开发

本章先对 Spring Boot 的 WebFlux（下文有时简称为 WebFlux）及其编程模型进行简单介绍，在此基础上介绍 Spring Boot 的 WebFlux 入门应用，以及在 WebFlux 开发中如何实现 RESTful 服务、如何访问 MongoDB 数据库、如何使用 Thymeleaf 和 MongoDB、如何访问 Redis 数据库、如何使用 WebSocket。

9.1 WebFlux 及其编程模型

9.1.1 WebFlux

Spring Boot 的 WebFlux 常用生产特性包括响应式 API、内嵌容器、Starter 组件、编程模型、适用性，还有对日志、Web、消息、测试及扩展等支持。

要了解 WebFlux，首先要了解什么是响应式流（Reactive Streams）。响应式流是 JVM 中面向流的库标准和规范，一般由发布者（发布元素到订阅者）、订阅者（消费元素）、订阅（发布者创建后与订阅者共享）、处理器（在发布者与订阅者之间处理数据）组成。响应流的特点包括：处理可能无限数量的元素；按顺序处理；组件之间异步传递；强制性非阻塞背压（Backpressure）。背压是一种常用策略，使得发布者拥有无限制的缓冲区存储元素，用于确保发布者发布元素太快时不会去压制订阅者。利用响应式流规范进行响应式编程是基于异步和事件驱动的非阻塞程序，响应式编程通过基于响应式流规范实现的框架 Reactor 去实现。

Spring Boot 的 WebFlux 就是基于 Reactor 实现的。Reactor 框架是 Spring Boot 中 WebFlux 响应库依赖，通响应式流并与其他响应库交互。它提供了两种响应式 API。响应式 API 一般是将 Publisher 作为输入，在框架内部转换成 Reactor 类型并处理逻辑，然后返回 Flux 或 Mono 作为输出。

Spring Boot 2.0 包括一个新的 spring-webflux 模块。该模块包含对响应式 HTTP 和 WebSocket 客户端的支持以及对 REST、HTML 和 WebSocket 交互等程序的支持。Spring 5 的 Web 模块中包含了 WebFlux 的 HTTP 抽象。类似于 Servlet API，WebFlux 提供了 WebHandler 的 API 去定义非阻塞 API 抽象接口。

WebFlux 默认是通过 Netty 启动，并且默认端口自动设置为 8080。另外还提供了对 Tomcat、Jetty、Undertow 等容器的支持。开发者添加依赖后，即可配置并使用对应的内嵌容器实例。

Spring Boot 的 Webflux 也提供了很多“开箱即用”的 Starter 组件。只需要在 Maven 配置文件 pom.xml 中添加对应的依赖配置，即可使用对应的 Starter 组件。例如，添加 spring-boot-starter-webflux 依赖，就可用于构建响应式 API 服务，其包含了 WebFlux 和内嵌容器等。

9.1.2 Spring Boot 的 WebFlux 编程模型

Spring Boot 的 WebFlux 开发有两种编程模型：一种是注解方式，示例代码如例 9-1 所示；另一种是使用功能性端点方式，示例代码如例 9-2 所示。

【例 9-1】 用注解方式进行 Spring Boot 的 WebFlux 开发的代码示例。

```
@RestController
@RequestMapping("/users")
public class MyRestController {
    @GetMapping("/{user}")
    public Mono<User> getUser(@PathVariable Long user) {
        //...
    }
    @GetMapping("/{user}/customers")
    public Flux<Customer> getUserCustomers(@PathVariable Long user) {
        //...
    }
    @DeleteMapping("/{user}")
    public Mono<User> deleteUser(@PathVariable Long user) {
        //...
    }
}
```

例 9-2 是例 9-1 的功能变体，它将路由配置与请求的实际处理分开，基于 lambda 轻量级编程模型来实现路由和处理请求。

【例 9-2】 使用功能性端点方式进行 Spring Boot 的 WebFlux 开发的代码示例。

```
//路由配置类
@Configuration
public class RoutingConfiguration {
    @Bean
```

```
    public RouterFunction<ServerResponse> monoRouterFunction(UserHandler
    userHandler) {
        return route(GET("/{user}").and(accept(APPLICATION_JSON)),
        userHandler::getUser)
.andRoute(GET("/{user}/customers").and(accept(APPLICATION_JSON)),
userHandler::getUserCustomers)
 .andRoute(DELETE("/{user}").and(accept(APPLICATION_JSON)),
 userHandler::deleteUser);
    }
}
//请求的实际处理类
@Component
public class UserHandler {
    public Mono<ServerResponse> getUser(ServerRequest request) {
        //...
    }
    public Mono<ServerResponse> getUserCustomers(ServerRequest request) {
        //...
    }
    public Mono<ServerResponse> deleteUser(ServerRequest request) {
        //...
    }
}
```

Spring Boot 的 WebFlux 和 Spring MVC 既有交集也有差异，如 Spring Boot 的 WebFlux 能更好地支撑并发。一般来说，Spring MVC 用于同步处理，Spring Boot 的 WebFlux 用于异步处理。如果用 Spring MVC 就能满足开发需要时，就没有必要使用 Spring Boot 的 WebFlux。在使用微服务体系结构时，Spring Boot 的 WebFlux 和 Spring MVC 可以混合使用。尤其是开发输入输出密集型服务时，可以选择 Spring Boot 的 WebFlux 去实现。

9.2 WebFlux 入门应用

视频讲解

9.2.1 添加依赖

在 pom.xml 文件中<dependencies>和</dependencies>之间添加所需的 Reactive Web 依赖，代码如例 9-3 所示。

【例 9-3】 添加 Reactive Web 依赖的代码示例。

```
<dependency>
        <groupId>org.springframework.boot</groupId>
        <artifactId>spring-boot-starter-webflux</artifactId>
</dependency>
<dependency>
        <groupId>io.projectreactor</groupId>
```

```
            <artifactId>reactor-test</artifactId>
            <scope>test</scope>
</dependency>
```

9.2.2 创建类 CityHandler

创建类 CityHandler，代码如例 9-4 所示。

【例 9-4】 创建类 CityHandler 的代码示例。

```
package com.bookcode.handler;
import org.springframework.http.MediaType;
import org.springframework.stereotype.Component;
import org.springframework.web.reactive.function.BodyInserters;
import org.springframework.web.reactive.function.server.ServerRequest;
import org.springframework.web.reactive.function.server.ServerResponse;
import reactor.core.publisher.Mono;
@Component
public class CityHandler {
    public Mono<ServerResponse> helloCity(ServerRequest request) {
        return ServerResponse.ok().contentType(MediaType.TEXT_PLAIN)
                .body(BodyInserters.fromObject("Hello City!"));
    }
}
```

9.2.3 创建类 CityRouter

创建类 CityRouter，代码如例 9-5 所示。

【例 9-5】 创建类 CityRouter 的代码示例。

```
package com.bookcode.router;
import com.bookcode.handler.CityHandler;
import org.springframework.context.annotation.Bean;
import org.springframework.context.annotation.Configuration;
import org.springframework.http.MediaType;
import org.springframework.web.reactive.function.server.RequestPredicates;
import org.springframework.web.reactive.function.server.RouterFunction;
import org.springframework.web.reactive.function.server.RouterFunctions;
import org.springframework.web.reactive.function.server.ServerResponse;
@Configuration
public class CityRouter {
    @Bean
    public RouterFunction<ServerResponse> routeCity(CityHandler cityHandler) {
        return RouterFunctions
                .route(RequestPredicates.GET("/hello")
                        .and(RequestPredicates.accept(MediaType.TEXT_
```

```
                        PLAIN)),
                    cityHandler::helloCity);
    }
}
```

9.2.4　运行程序

运行程序，可以在控制台中看到服务器 Netty 成功启动。在浏览器中输入 localhost:8080/hello，结果如图 9-1 所示。

图 9-1　在浏览器中输入 localhost:8080/hello 后的结果

9.3　实现基于 WebFlux 的 RESTful 服务

视频讲解

9.3.1　添加依赖

在项目 pom.xml 文件中<dependencies>和</dependencies>之间添加 Reactive Web 依赖，代码如例 9-3 所示。

9.3.2　创建类 User

在包 com.bookcode.entity 中创建类 User，代码如例 9-6 所示。

【例 9-6】 创建类 User 的代码示例。

```
package com.bookcode.entity;
public class User {
    private long id;
    private String firstname;
    private String lastname;
    private int age;
    public User() {    }
    public User(long id, String firstname, String lastname, int age) {
        this.id = id;
        this.firstname = firstname;
        this.lastname = lastname;
        this.age = age;
```

```
    }
    public long getId() {
        return id;
    }
    public void setId(Long id) {
        this.id = id;
    }
    public String getFirstname() {
        return firstname;
    }
    public void setFirstname(String firstname) {
        this.firstname = firstname;
    }
    public String getLastname() {
        return lastname;
    }
    public void setLastname(String lastname) {
        this.lastname = lastname;
    }
    public int getAge() {
        return age;
    }
    public void setAge(int age) {
        this.age = age;
    }
    @Override
    public String toString() {
        String info = String.format("id = %d, firstname = %s, lastname = %s,
        age = %d", id, firstname, lastname, age);
        return info;
    }
}
```

9.3.3 创建类 UserController

在 com.bookcode.controller 包中创建类 UserController，代码如例 9-7 所示。

【例 9-7】 创建类 UserController 的代码示例。

```
package com.bookcode.controller;
import com.bookcode.entity.User;
import org.slf4j.Logger;
import org.slf4j.LoggerFactory;
import org.springframework.http.HttpStatus;
import org.springframework.http.ResponseEntity;
import org.springframework.web.bind.annotation.*;
import reactor.core.publisher.Flux;
```

```
import reactor.core.publisher.Mono;
import javax.annotation.PostConstruct;
import java.util.HashMap;
import java.util.Map;
import java.util.stream.Collectors;@RestController
@RequestMapping(path = "/api/user")
public class UserController {
    private static final Logger logger = LoggerFactory.getLogger
    (UserController.class);
    Map<Long, User> users = new HashMap<>();
    @PostConstruct
    public void init() throws Exception {
        users.put(Long.valueOf(1), new User(1, "Jack", "Smith", 20));
        users.put(Long.valueOf(2), new User(2, "Peter", "Johnson", 25));
    }
    @GetMapping("/index")
    public Flux<User> getAll() {
        return Flux.fromIterable(users.entrySet().stream()
                .map(entry -> entry.getValue())
                .collect(Collectors.toList()));
    }
    //获取单个用户
    @GetMapping("/{id}")
    public Mono<User> getCustomer(@PathVariable Long id) {
        return Mono.justOrEmpty(users.get(id));
    }
    //创建用户
    @PostMapping("/post")
    public Mono<ResponseEntity<String>> postUser(@RequestBody User user) {
        users.put(user.getId(), user);
        logger.info("########### POST:" + user);
        return Mono.just(new ResponseEntity<>("Post Successfully!",
        HttpStatus.CREATED));
    }
    //修改用户
    @PutMapping("/put/{id}")
    public Mono<ResponseEntity<User>> putCustomer(@PathVariable Long id,
    @RequestBody User user) {
        user.setId(id);
        users.put(id, user);
        System.out.println("########### PUT:" + user);
        return Mono.just(new ResponseEntity<>(user, HttpStatus.CREATED));
    }
    @DeleteMapping("/delete/{id}")
    public Mono<ResponseEntity<String>> deleteMethod(@PathVariable Long id) {
```

```
        users.remove(id);
        return Mono.just(new ResponseEntity<>("Delete Successfully!",
HttpStatus.ACCEPTED));
    }
}
```

9.3.4 运行程序

运行程序，在浏览器中输入 localhost:8080/api/user/index，显示所有 User 实例，结果如图 9-2 所示。在 Postman 工具中显示所有 User 实例的结果类似，如图 9-3 所示。在 Postman 工具中增加 firstname、lastname 都为 z 的 User 实例，结果如图 9-4 所示。注意，选择的数据格式是 JSON(application/json)。增加 User 实例后在 Postman 工具中显示所有 User 实例的结果如图 9-5 所示。在浏览器中显示第 2 位 User 实例信息的结果如图 9-6 所示。

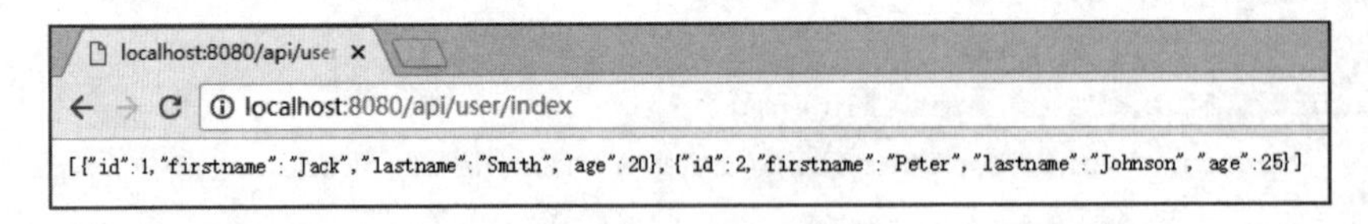

图 9-2 在浏览器中输入 localhost:8080/api/user/index 后的结果

图 9-3 在工具 Postman 中显示所有 User 实例的结果

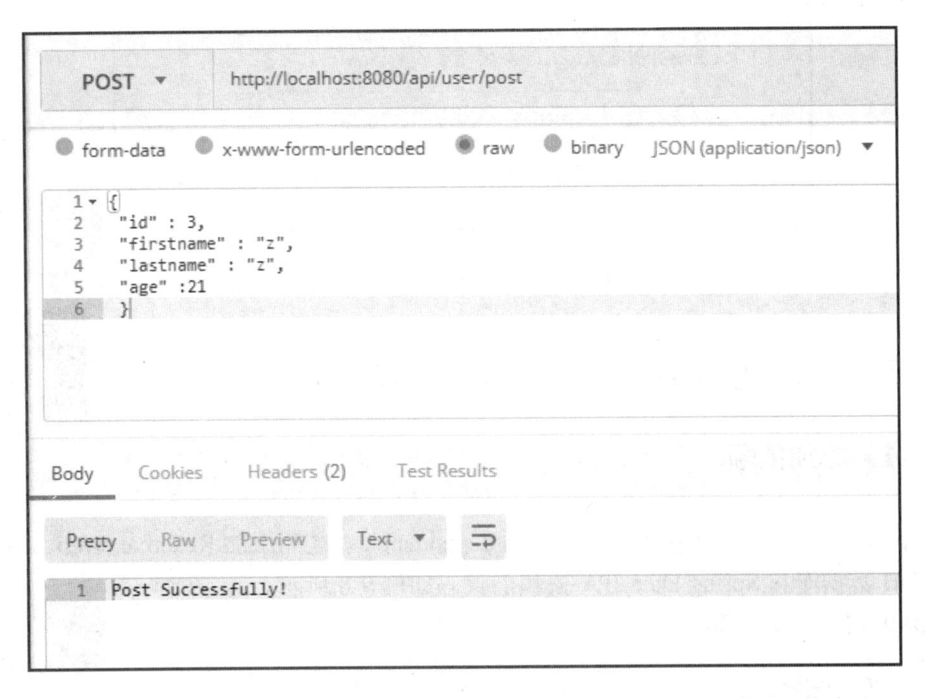

图 9-4 在 Postman 工具中增加 firstname、lastname 都为 z 的 User 实例后的结果

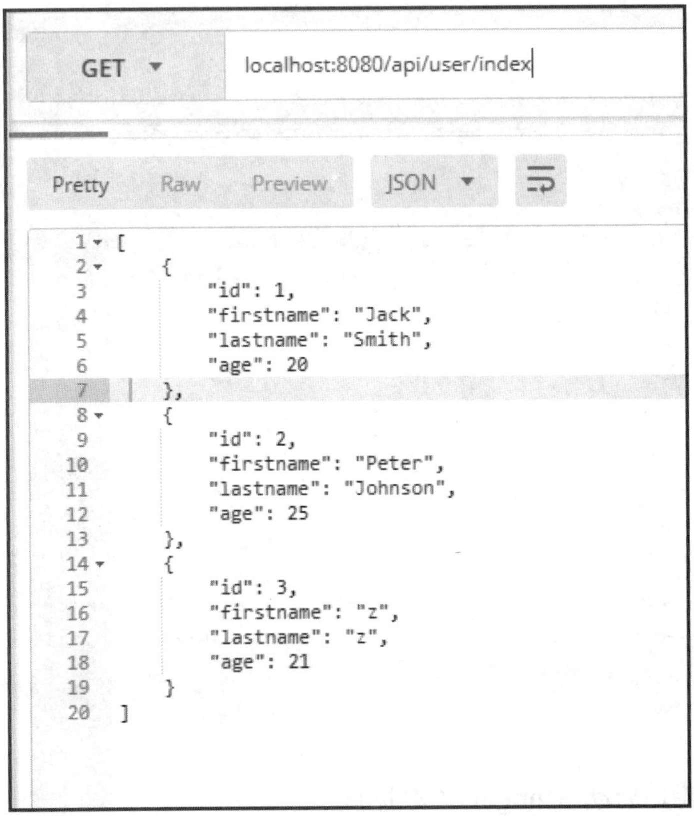

GET localhost:8080/api/user/index

Pretty Raw Preview JSON

```
[
    {
        "id": 1,
        "firstname": "Jack",
        "lastname": "Smith",
        "age": 20
    },
    {
        "id": 2,
        "firstname": "Peter",
        "lastname": "Johnson",
        "age": 25
    },
    {
        "id": 3,
        "firstname": "z",
        "lastname": "z",
        "age": 21
    }
]
```

图 9-5 增加 User 实例后在 Postman 工具中显示所有 User 实例的结果

图 9-6 在浏览器中显示第 2 位 User 实例信息的结果

9.4 基于 WebFlux 访问 MongoDB 数据库

视频讲解

9.4.1 添加依赖

在 pom.xml 文件中<dependencies>和</dependencies>之间添加 Reactive Web、Reactive MongoDB 数据库和 Spring Data JPA 依赖，代码如例 9-8 所示。

【例 9-8】 添加依赖的代码示例。

```
<dependency>
        <groupId>org.springframework.boot</groupId>
        <artifactId>spring-boot-starter-webflux</artifactId>
</dependency>
<dependency>
        <groupId>org.springframework.boot</groupId>
        <artifactId>spring-boot-starter-data-mongodb-reactive</artifactId>
</dependency>
<dependency>
        <groupId>org.springframework.boot</groupId>
        <artifactId>spring-boot-starter-test</artifactId>
        <scope>test</scope>
</dependency>
<dependency>
        <groupId>io.projectreactor</groupId>
        <artifactId>reactor-test</artifactId>
        <scope>test</scope>
</dependency>
<dependency>
        <groupId>org.springframework.boot</groupId>
        <artifactId>spring-boot-starter-data-jpa</artifactId>
        <scope>test</scope>
</dependency>
```

9.4.2 安装并启动 MongoDB 数据库

如同安装一般软件一样安装 MongoDB 数据库。安装好 MongoDB 数据库之后，可以打开 Windows 命令处理程序 CMD，输入如例 9-9 所示的命令启动 MongoDB 数据库服务。

【例 9-9】 启动 MongoDB 数据库服务的命令代码示例。

```
mongod --dbpath "D:\mongodb\data\db"
```

9.4.3　创建类 Person

在包 com.bookcode.entity 中创建类 Person，代码如例 9-10 所示。

【例 9-10】 创建类 Person 的代码示例。

```
package com.bookcode.entity;
import org.springframework.data.annotation.Id;
public class Person {
    @Id
    public String id;
    private String username;
    public String getId() {
        return id;
    }
    public void setId(String id) {
        this.id = id;
    }
    public String getUsername() {
        return username;
    }
    public void setUsername(String username) {
        this.username = username;
    }
}
```

9.4.4　创建接口 PersonRepository

在包 com.bookcode.dao 中创建接口 PersonRepository，代码如例 9-11 所示。

【例 9-11】 创建接口 PersonRepository 的代码示例。

```
package com.bookcode.dao;
import com.bookcode.entity.Person;
import org.springframework.data.mongodb.repository.ReactiveMongoRepository;
import org.springframework.stereotype.Repository;
@Repository
public interface PersonRepository extends ReactiveMongoRepository<Person,
String> {
}
```

9.4.5 创建类 PersonController

在包 com.bookcode.controller 中创建类 PersonController，代码如例 9-12 所示。

【例 9-12】 创建类 PersonController 的代码示例。

```
package com.bookcode.controller;
import com.bookcode.dao.PersonRepository;
import com.bookcode.entity.Person;
import org.reactivestreams.Publisher;
import org.springframework.web.bind.annotation.*;
import reactor.core.publisher.Flux;
import reactor.core.publisher.Mono;
@RestController
public class PersonController {
    private final PersonRepository personRepository;
    public PersonController(PersonRepository personRepository) {
        this.personRepository = personRepository;
    }
    //正常 MVC 模式
    @GetMapping("/")
    public String hello(){
        return "hello!";
    }
    //新增一个 Person
    @PostMapping("/person")
    public Mono<Void> add(@RequestBody Publisher<Person> person){
        return personRepository.insert(person).then();
    }
    @GetMapping("/person/{id}")
    public Mono<Person> getById(@PathVariable Long id){
        return personRepository.findById(id);
    }
    @GetMapping("/person/list")
    public Flux<Person> list(){
        return personRepository.findAll();
    }
    @DeleteMapping("/person/{id}")
    public Mono<Void> delete(@PathVariable Long id){
        return personRepository.deleteById(id).then();
    }
}
```

9.4.6 修改配置文件 application.properties

修改配置文件 application.properties，代码如例 9-13 所示。

【例 9-13】 修改配置文件 application.properties 的代码示例。

```
spring.data.mongodb.host=localhost
spring.data.mongodb.port=27017
```

9.4.7　运行程序

启动 MongoDB 数据库后，运行程序，就可以用工具 Postman 来增加 person 记录到数据库中，如图 9-7 所示。与此同时，MongoDB 数据库中 test 自动生成了记录，结果在可视化工具 Robo 3T 中的显示情况如图 9-8 所示。

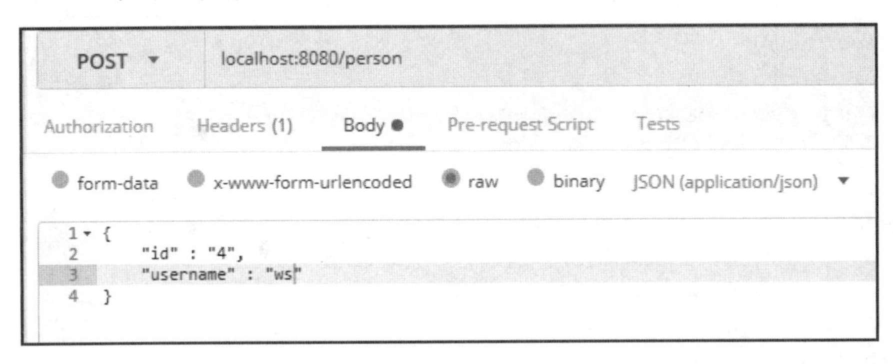

图 9-7　用工具 Postman 来增加 person 记录

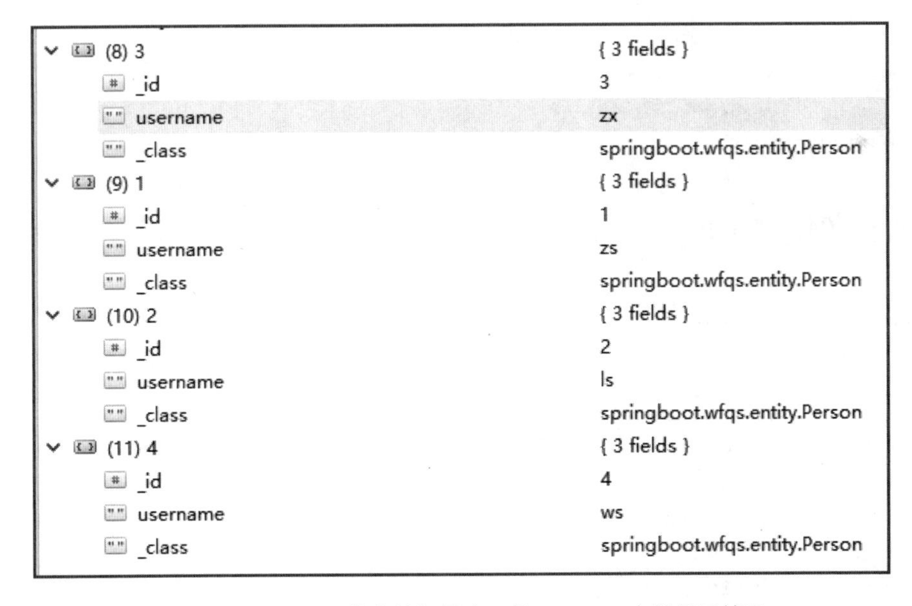

图 9-8　MongoDB 的存储记录在工具 Robo 3T 中的显示情况

9.5　基于 WebFlux 使用 Thymeleaf 和 MongoDB

视频讲解

9.5.1　添加依赖

在 pom.xml 文件中<dependencies>和</dependencies>之间添加依赖，代码如例 9-14

所示。

【例 9-14】 添加依赖的代码示例。

```
<dependency>
          <groupId>org.springframework.boot</groupId>
          <artifactId>spring-boot-starter-thymeleaf</artifactId>
</dependency>
<dependency>
          <groupId>org.springframework.boot</groupId>
          <artifactId>spring-boot-starter-webflux</artifactId>
</dependency>
<dependency>
          <groupId>org.springframework.boot</groupId>
          <artifactId>spring-boot-starter-data-mongodb-reactive</artifactId>
</dependency>
<dependency>
          <groupId>org.springframework.boot</groupId>
          <artifactId>spring-boot-starter-test</artifactId>
          <scope>test</scope>
</dependency>
<dependency>
          <groupId>io.projectreactor</groupId>
          <artifactId>reactor-test</artifactId>
          <scope>test</scope>
</dependency>
```

9.5.2 创建类 City

创建类 City，代码如例 9-15 所示。

【例 9-15】 创建类 City 的代码示例。

```
package com.bookcode.entity;
public class City {
   private Long id;              //城市编号
   private Long provinceId;      //省份编号
   private String cityName;
   private String description;
   public Long getId() {
      return id;
   }
   public void setId(Long id) {
      this.id = id;
   }
   public Long getProvinceId() {
      return provinceId;
```

```
    }
    public void setProvinceId(Long provinceId) {
        this.provinceId = provinceId;
    }
    public String getCityName() {
        return cityName;
    }
    public void setCityName(String cityName) {
        this.cityName = cityName;
    }
    public String getDescription() {
        return description;
    }
    public void setDescription(String description) {
        this.description = description;
    }
}
```

9.5.3　创建接口 CityRepository

在包 com.bookcode.dao 中创建接口 CityRepository，代码如例 9-16 所示。

【例 9-16】 创建接口 CityRepository 的代码示例。

```
package com.bookcode.dao;
import com.bookcode.entity.City;
import org.springframework.data.mongodb.repository.ReactiveMongoRepository;
import org.springframework.stereotype.Repository;
import reactor.core.publisher.Mono;
@Repository
public interface CityRepository extends ReactiveMongoRepository<City, Long> {
    Mono<City> findByCityName(String cityName);
}
```

9.5.4　创建类 CityHandler

创建类 CityHandler，代码如例 9-17 所示。

【例 9-17】 创建类 CityHandler 的代码示例。

```
package com.bookcode.handler;
import com.bookcode.dao.CityRepository;
import com.bookcode.entity.City;
import org.springframework.beans.factory.annotation.Autowired;
import org.springframework.stereotype.Component;
import reactor.core.publisher.Flux;
import reactor.core.publisher.Mono;
```

```
@Component
public class CityHandler {
    private final CityRepository cityRepository;
    @Autowired
    public CityHandler(CityRepository cityRepository) {
        this.cityRepository = cityRepository;
    }
    public Mono<City> save(City city) {
        return cityRepository.save(city);
    }
    public Mono<City> findCityById(Long id) {
        return cityRepository.findById(id);
    }
    public Flux<City> findAllCity() {
        return cityRepository.findAll();
    }
    public Mono<City> modifyCity(City city) {
        return cityRepository.save(city);
    }
    public Mono<Long> deleteCity(Long id) {
        cityRepository.deleteById(id);
        return Mono.create(cityMonoSink -> cityMonoSink.success(id));
    }
    public Mono<City> getByCityName(String cityName) {
        return cityRepository.findByCityName(cityName);
    }
}
```

9.5.5 创建类 CityController

创建类 CityController，其代码如例 9-18 所示。

【例 9-18】 创建类 CityController 的代码示例。

```
package com.bookcode.controller;
import com.bookcode.entity.City;
import com.bookcode.handler.CityHandler;
import org.springframework.beans.factory.annotation.Autowired;
import org.springframework.stereotype.Controller;
import org.springframework.ui.Model;
import org.springframework.web.bind.annotation.*;
import reactor.core.publisher.Flux;
import reactor.core.publisher.Mono;
@Controller
@RequestMapping(value = "/city")
public class CityController {
```

```
    @Autowired
    private CityHandler cityHandler;
    @GetMapping(value = "/{id}")
    @ResponseBody
    public Mono<City> findCityById(@PathVariable("id") Long id) {
        return cityHandler.findCityById(id);
    }
    @GetMapping()
    @ResponseBody
    public Flux<City> findAllCity() {
        return cityHandler.findAllCity();
    }
    @PostMapping()
    @ResponseBody
    public Mono<City> saveCity(@RequestBody City city) {
        return cityHandler.save(city);
    }
    @PutMapping()
    @ResponseBody
    public Mono<City> modifyCity(@RequestBody City city) {
        return cityHandler.modifyCity(city);
    }
    @DeleteMapping(value = "/{id}")
    @ResponseBody
    public Mono<Long> deleteCity(@PathVariable("id") Long id) {
        return cityHandler.deleteCity(id);
    }
    private static final String CITY_LIST_PATH_NAME = "cityList";
    private static final String CITY_PATH_NAME = "city";
    @GetMapping("/page/list")
    public String listPage(final Model model) {
        final Flux<City> cityFluxList = cityHandler.findAllCity();
        model.addAttribute("cityList", cityFluxList);
        return CITY_LIST_PATH_NAME;
    }
    @GetMapping("/getByName")
    public String getByCityName(final Model model,
     @RequestParam(value="cityName",required = false, defaultValue ="xuzhou")
     String cityName) {
        final Mono<City> city = cityHandler.getByCityName(cityName);
        model.addAttribute("city", city);
        return CITY_PATH_NAME;
    }
}
```

9.5.6 创建文件 cityList.html

在 src/main/resources/templates 目录下创建文件 cityList.html，代码如例 9-19 所示。

【例 9-19】 创建文件 cityList.html 的代码示例。

```
<!DOCTYPE html>
<html lang="zh-CN" xmlns:th="http://www.thymeleaf.org">
<head>
    <meta charset="UTF-8"/>
    <title>城市列表</title>
</head>
<body>
<div>
    <table>
        <legend>
            <strong>城市列表</strong>
        </legend>
        <thead>
        <tr>
            <th>城市编号</th>
            <th>省份编号</th>
            <th>名称</th>
            <th>描述</th>
        </tr>
        </thead>
        <tbody>
        <tr th:each="city : ${cityList}">
            <td th:text="${city.id}"></td>
            <td th:text="${city.provinceId}"></td>
            <td th:text="${city.cityName}"></td>
            <td th:text="${city.description}"></td>
        </tr>
        </tbody>
    </table>
</div>
</body>
</html>
```

9.5.7 创建文件 city.html

在 src/main/resources/templates 目录下创建文件 city.html，代码如例 9-20 所示。

【例 9-20】 创建文件 city.html 的代码示例。

```
<!DOCTYPE html>
<html lang="zh-CN" xmlns:th="http://www.thymeleaf.org">
```

```
<head>
    <meta charset="UTF-8"/>
    <title>城市</title>
</head>
<body>
<div>
    <table>
        <legend>
            <strong>城市单个查询</strong>
        </legend>
        <tbody>
            <td th:text="${city.id}"></td>
            <td th:text="${city.provinceId}"></td>
            <td th:text="${city.cityName}"></td>
            <td th:text="${city.description}"></td>
        </tbody>
    </table>
</div>
</body>
</html>
```

9.5.8　运行程序

修改配置文件 application.properties，代码如例 9-13 所示。

运行程序后，启动 MongoDB，在浏览器中输入 localhost:8080/city/page/list 后，在浏览器中显示所有 city 记录信息，结果如图 9-9 所示。为了利用 cityName 查询 xuzhou（徐州）信息，在浏览器中输入 http://localhost:8080/city/getByName?cityName=xuzhou，结果如图 9-10 所示。

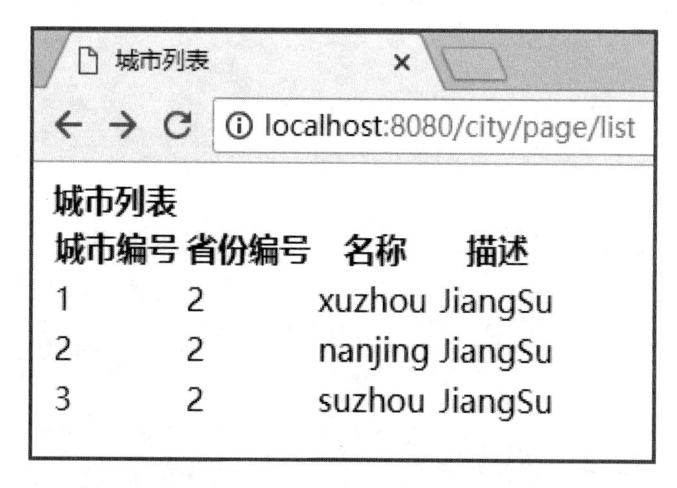

图 9-9　在浏览器中输入 localhost:8080/city/page/list 后的结果

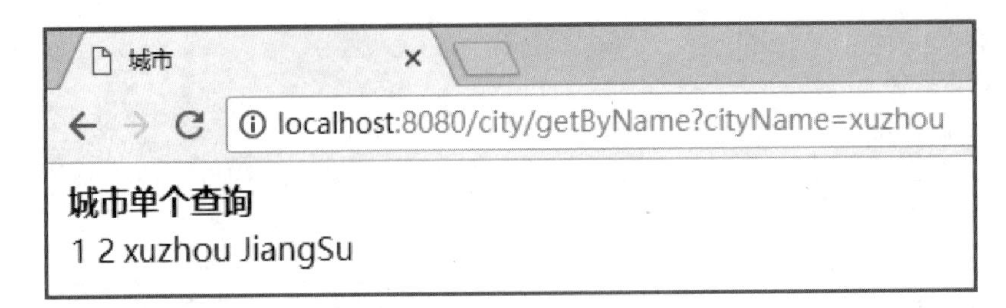

图 9-10　在浏览器中输入 http://localhost:8080/city/getByName?cityName=xuzhou 后的结果

9.6　基于 WebFlux 访问 Redis 数据库

视频讲解

Redis 是一种可以持久存储的缓存系统，是用 C 语言编写、支持网络、可基于内存并可持久化的日志型 Key-Value（键-值）数据库，并提供多种语言的 API。Redis 支持存储的值类型很多，包括字符串、链表、集合、有序集合和哈希类型。在此基础上，Redis 支持各种不同方式的排序。为了保证效率，数据都是缓存在内存中。它提供了 Java、C/C++、Python 等客户端，使用很方便。Redis 支持主从同步，从服务器上的数据可以和主服务器上的数据同步。同步对读取操作的可扩展性和数据冗余很有帮助。

9.6.1　添加依赖

在 pom.xml 文件中<dependencies>和</dependencies>之间添加依赖，代码如例 9-21 所示。

【例 9-21】 添加依赖的代码示例。

```
<dependency>
          <groupId>org.springframework.boot</groupId>
          <artifactId>spring-boot-starter-data-redis-reactive</artifactId>
</dependency>
 <dependency>
          <groupId>org.springframework.boot</groupId>
          <artifactId>spring-boot-starter-webflux</artifactId>
</dependency>
<dependency>
          <groupId>org.springframework.boot</groupId>
          <artifactId>spring-boot-starter-test</artifactId>
          <scope>test</scope>
</dependency>
<dependency>
          <groupId>io.projectreactor</groupId>
          <artifactId>reactor-test</artifactId>
          <scope>test</scope>
</dependency>
```

9.6.2　创建类 Coffee

在包 com.bookcode.entity 中创建类 Coffee，代码如例 9-22 所示。

【例 9-22】 创建类 Coffee 的代码示例。

```
package com.bookcode.entity;
public class Coffee {
    private String id;
    private String name;
    public String getId() {
        return id;
    }
    public void setId(String id) {
        this.id = id;
    }
    public String getName() {
        return name;
    }
    public void setName(String name) {
        this.name = name;
    }
    public Coffee()   {   }
    public Coffee(String id,String name){
        this.id=id;
        this.name=name;
    }
}
```

9.6.3　创建类 CoffeeConfiguration

在包 com.bookcode.config 中创建类 CoffeeConfiguration，代码如例 9-23 所示。

【例 9-23】 创建类 CoffeeConfiguration 的代码示例。

```
package com.bookcode.config;
import com.bookcode.entity.Coffee;
import org.springframework.context.annotation.Bean;
import org.springframework.context.annotation.Configuration;
import org.springframework.data.redis.connection.ReactiveRedisConnectionFactory;
import org.springframework.data.redis.core.ReactiveRedisOperations;
import org.springframework.data.redis.core.ReactiveRedisTemplate;
import org.springframework.data.redis.serializer.Jackson2JsonRedisSerializer;
import org.springframework.data.redis.serializer.RedisSerializationContext;
import org.springframework.data.redis.serializer.StringRedisSerializer;
@Configuration       //配置类
```

```
public class CoffeeConfiguration {
    @Bean
 ReactiveRedisOperations<String, Coffee> redisOperations
 (ReactiveRedisConnectionFactory factory) {
 //Redis 没有表结构的概念，需要使用 JSON 格式的文本作为 Redis 和 Java 普通对象互相交换
 //数据的存储格式
 //RedisTemplate 初始化
 //为了正确调用 RedisTemplate，必须对其进行一些初始化工作，即主要对它存取的字符串进
 //行一个 JSON 格式的系列化初始配置
  Jackson2JsonRedisSerializer<Coffee> serializer = new
  Jackson2JsonRedisSerializer<>(Coffee.class);
        RedisSerializationContext.RedisSerializationContextBuilder<String,
        Coffee> builder = RedisSerializationContext.newSerializationContext(new
                StringRedisSerializer());
        RedisSerializationContext<String, Coffee> context = builder.value
        (serializer).build();
        return new ReactiveRedisTemplate<>(factory, context);
    }
}
```

9.6.4 创建类 CoffeeLoader

在包 com.bookcode.loader 中创建类 CoffeeLoader，代码如例 9-24 所示。

【例 9-24】 创建类 CoffeeLoader 的代码示例。

```
package com.bookcode.loader;
import com.bookcode.entity.Coffee;
import org.springframework.data.redis.connection.ReactiveRedisConnectionFactory;
import org.springframework.data.redis.core.ReactiveRedisOperations;
import org.springframework.stereotype.Component;
import reactor.core.publisher.Flux;
import javax.annotation.PostConstruct;
import java.util.UUID;
@Component              //组件
public class CoffeeLoader {
    private final ReactiveRedisConnectionFactory factory;
    private final ReactiveRedisOperations<String, Coffee> coffeeOps;
    public CoffeeLoader(ReactiveRedisConnectionFactory factory,
    ReactiveRedisOperations<String, Coffee> coffeeOps) {
        this.factory = factory;
        this.coffeeOps = coffeeOps;    }
    @PostConstruct
    public void loadData() {
```

```
        factory.getReactiveConnection().serverCommands().flushAll().thenMany(
                Flux.just("Jet Black Redis", "Darth Redis", "Black Alert
               Redis")
                        .map(name -> new Coffee(UUID.randomUUID().toString(),
                         name))
                        .flatMap(coffee -> coffeeOps.opsForValue().set
                        (coffee.getId(), coffee)))
                .thenMany(coffeeOps.keys("*")
                        .flatMap(coffeeOps.opsForValue()::get))
                .subscribe(System.out::println);
    }
}
```

9.6.5 运行程序

运行程序，可以在工具 Redis Desktop Manager 中观察到增加了三条记录，结果如图 9-11 所示。

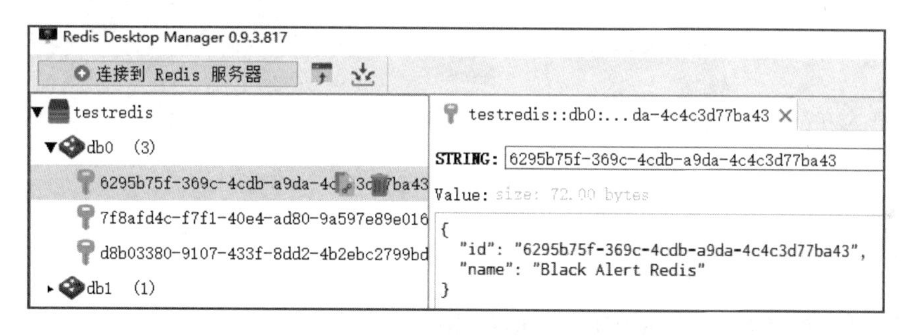

图 9-11 在工具 Redis Desktop Manager 中显示的结果

9.6.6 创建类 City

在包 com.bookcode.entity 中创建类 City，代码如例 9-25 所示。

【例 9-25】 创建类 City 的代码示例。

```
package com.bookcode.entity;
import org.springframework.data.annotation.Id;
import java.io.Serializable;
public class City implements Serializable {
    @Id
    private Long id;
    private Long provinceId;
    private String cityName;
    private String description;
```

```
    public Long getId() {
        return id;
    }
    public void setId(Long id) {
        this.id = id;
    }
    public Long getProvinceId() {
        return provinceId;
    }
    public void setProvinceId(Long provinceId) {
        this.provinceId = provinceId;
    }
    public String getCityName() {
        return cityName;
    }
    public void setCityName(String cityName) {
        this.cityName = cityName;
    }
    public String getDescription() {
        return description;
    }
    public void setDescription(String description) {
        this.description = description;
    }
}
```

9.6.7 创建类 CityWebFluxController

创建类 CityWebFluxController，代码如例 9-26 所示。

【例 9-26】 创建类 CityWebFluxController 的代码示例。

```
package com.bookcode.controller;
import com.bookcode.entity.City;
import org.springframework.beans.factory.annotation.Autowired;
import org.springframework.data.redis.core.RedisTemplate;
import org.springframework.data.redis.core.ValueOperations;
import org.springframework.web.bind.annotation.*;
import reactor.core.publisher.Mono;
import java.util.concurrent.TimeUnit;
@RestController
@RequestMapping(value = "/city")
public class CityWebFluxController {
    @Autowired
    private RedisTemplate redisTemplate;
```

```
    @GetMapping(value = "/{id}")
    public Mono<City> findCityById(@PathVariable("id") Long id) {
        String key = "city_" + id;
        ValueOperations<String, City> operations = redisTemplate.opsForValue();
        boolean hasKey = redisTemplate.hasKey(key);
        City city = operations.get(key);
        if (!hasKey) {
            return Mono.create(monoSink -> monoSink.success(null));
        }
        return Mono.create(monoSink -> monoSink.success(city));
    }
    @PostMapping()
    public Mono<City> saveCity(@RequestBody City city) {
        String key = "city_" + city.getId();
        ValueOperations<String, City> operations = redisTemplate.
        opsForValue();
        operations.set(key, city, 60, TimeUnit.SECONDS);
        return Mono.create(monoSink -> monoSink.success(city));
    }
    @DeleteMapping(value = "/{id}")
    public Mono<Long> deleteCity(@PathVariable("id") Long id) {
        String key = "city_" + id;
        boolean hasKey = redisTemplate.hasKey(key);
        if (hasKey) {
            redisTemplate.delete(key);
        }
        return Mono.create(monoSink -> monoSink.success(id));
    }
}
```

9.6.8　修改配置文件 application.properties

修改配置文件 application.properties，代码如例 9-27 所示。

【例 9-27】　修改配置文件 application.properties 的代码示例。

```
#设置 Redis 数据库信息
spring.redis.host=localhost
spring.redis.port=6379
```

9.6.9　运行程序

运行程序，在工具 Postman 中添加一条城市信息记录，结果如图 9-12 所示。在工具 Postman 中获取第 2 条城市信息记录，结果如图 9-13 所示。

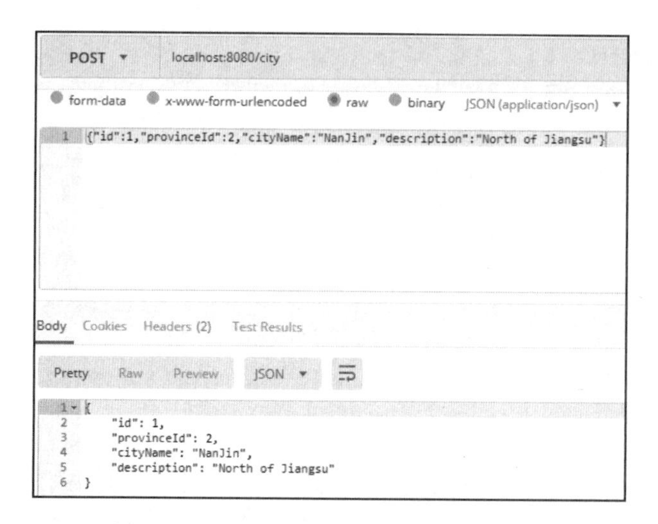

图 9-12　在工具 Postman 中添加一条城市信息记录的结果

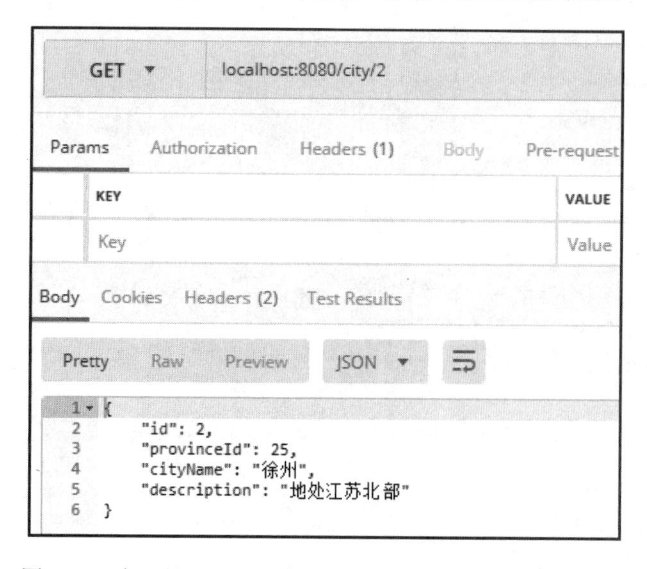

图 9-13　在工具 Postman 中获取第 2 条城市信息记录的结果

9.6.10　创建类 CityWebFluxReactiveController

创建类 CityWebFluxReactiveController，代码如例 9-28 所示，其功能与例 9-26 相同。

【例 9-28】 创建类 CityWebFluxReactiveController 的代码示例。

```
package com.bookcode.controller;
import com.bookcode.entity.City;
import org.springframework.beans.factory.annotation.Autowired;
import org.springframework.data.redis.core.ReactiveRedisTemplate;
import org.springframework.data.redis.core.ReactiveValueOperations;
import org.springframework.web.bind.annotation.*;
import reactor.core.publisher.Mono;
```

```
@RestController
@RequestMapping(value = "/city2")
public class CityWebFluxReactiveController {
    @Autowired
    private ReactiveRedisTemplate reactiveRedisTemplate;
    @GetMapping(value = "/{id}")
    public Mono<City> findCityById(@PathVariable("id") Long id) {
        String key = "city_" + id;
        ReactiveValueOperations<String, City> operations =
        reactiveRedisTemplate.opsForValue();
        Mono<City> city = operations.get(key);
        return city;
    }
    @PostMapping
    public Mono<City> saveCity(@RequestBody City city) {
        String key = "city_" + city.getId();
        ReactiveValueOperations<String, City> operations =
        reactiveRedisTemplate.opsForValue();
        return operations.getAndSet(key, city);
    }
    @DeleteMapping(value = "/{id}")
    public Mono<Long> deleteCity(@PathVariable("id") Long id) {
        String key = "city_" + id;
        return reactiveRedisTemplate.delete(key);
    }
}
```

9.7　基于 WebFlux 使用 WebSocket

视频讲解

WebSocket 协议是基于 TCP 的网络协议，它实现了浏览器与服务器全双工通信。在 WebSocket 中，浏览器和服务器只需要完成一次握手，就可以创建持久性的连接，浏览器和服务器之间就形成了一条快速通道。两者之间就直接可以互相传送数据。很多网站为了实现推送技术所用的技术都是 Ajax 轮询。轮询是在特定的的时间间隔（如 1 秒），由客户端的浏览器向服务器发出 HTTP 请求，然后由服务器返回最新的数据给浏览器。这种模式有明显的缺点，即浏览器要不断地向服务器发出请求。然而，HTTP 请求可能包含较长的头部，真正有效的数据可能只是很小的一部分，这样会浪费很多的带宽等资源。用 WebSocket 能更好地节省服务器资源和带宽，并且能够更实时地进行通信。

9.7.1　添加依赖

在 pom.xml 文件中<dependencies>和</dependencies>之间添加依赖，代码如例 9-29 所示。

【例 9-29】 添加依赖的代码示例。

```
<dependency>
        <groupId>org.springframework.boot</groupId>
        <artifactId>spring-boot-starter-webflux</artifactId>
</dependency>
<dependency>
        <groupId>org.springframework.boot</groupId>
        <artifactId>spring-boot-starter-test</artifactId>
        <scope>test</scope>
</dependency>
<dependency>
        <groupId>io.projectreactor</groupId>
        <artifactId>reactor-test</artifactId>
        <scope>test</scope>
</dependency>
```

9.7.2 创建类 EchoHandler

在包 com.bookcode.handler 中创建类 EchoHandler，代码如例 9-30 所示。

【例 9-30】 创建类 EchoHandler 的代码示例。

```
package com.bookcode.Handler;
import org.springframework.stereotype.Component;
import org.springframework.web.reactive.socket.WebSocketHandler;
import org.springframework.web.reactive.socket.WebSocketSession;
import reactor.core.publisher.Mono;
@Component
public class EchoHandler implements WebSocketHandler {
   @Override
   public Mono<Void> handle(final WebSocketSession session) {
      return session.send(
            session.receive()
                  .map(msg -> session.textMessage(
                        "服务端返回：小明，  " + msg.getPayloadAsText())));
   }
}
```

9.7.3 创建类 WebSocketConfiguration

在 com.bookcode.config 包中创建类 WebSocketConfiguration，代码如例 9-31 所示。

【例 9-31】 创建类 WebSocketConfiguration 的代码示例。

```
package com.bookcode.config;
import com.bookcode.Handler.EchoHandler;
```

```
import org.springframework.beans.factory.annotation.Autowired;
import org.springframework.context.annotation.Bean;
import org.springframework.context.annotation.Configuration;
import org.springframework.core.Ordered;
import org.springframework.web.reactive.HandlerMapping;
import org.springframework.web.reactive.handler.SimpleUrlHandlerMapping;
import org.springframework.web.reactive.socket.WebSocketHandler;
import org.springframework.web.reactive.socket.server.support.
WebSocketHandlerAdapter;
import java.util.HashMap;
import java.util.Map;
@Configuration
public class WebSocketConfiguration {
    @Autowired
    @Bean
    public HandlerMapping webSocketMapping(final EchoHandler echoHandler) {
        final Map<String, WebSocketHandler> map = new HashMap<>();
        map.put("/echo", echoHandler);
        final SimpleUrlHandlerMapping mapping = new
        SimpleUrlHandlerMapping();
        mapping.setOrder(Ordered.HIGHEST_PRECEDENCE);
        mapping.setUrlMap(map);
        return mapping;
    }
    @Bean
    public WebSocketHandlerAdapter handlerAdapter() {
        return new WebSocketHandlerAdapter();
    }
}
```

9.7.4　创建类 WSClient

在 test/Java/com.bookcode 目录下创建类 WSClient，代码如例 9-32 所示。

【例 9-32】 创建类 WSClient 的代码示例。

```
package com.bookcode;
import org.springframework.web.reactive.socket.WebSocketMessage;
import org.springframework.web.reactive.socket.client.
ReactorNettyWebSocketClient;
import org.springframework.web.reactive.socket.client.WebSocketClient;
import reactor.core.publisher.Flux;
import java.net.URI;
import java.time.Duration;
public class WSClient {
    public static void main(final String[] args) {
```

```
        final WebSocketClient client = new ReactorNettyWebSocketClient();
        client.execute(URI.create("ws://localhost:8080/echo"), session ->
                session.send(Flux.just(session.textMessage("你好")))
                .thenMany(session.receive().take(1).map(WebSocketMessage::
                getPayloadAsText))
                        .doOnNext(System.out::println)
                        .then())
                .block(Duration.ofMillis(5000));
    }
}
```

9.7.5 创建文件 websocket-client.html

在 resources/static 目录下创建文件 websocket-client.html，代码如例 9-33 所示。

【例 9-33】 创建文件 websocket-client.html 的代码示例。

```
<!DOCTYPE html>
<html lang="en">
<head>
  <meta charset="UTF-8"/>
  <title>Client WebSocket</title>
</head>
<body>
<div class="chat"></div>
<script>
  var clientWebSocket = new WebSocket("ws://localhost:8080/echo");
  clientWebSocket.onopen = function () {
    console.log("clientWebSocket.onopen", clientWebSocket);
    console.log("clientWebSocket.readyState", "websocketstatus");
    clientWebSocket.send("你好！");
  }
  clientWebSocket.onclose = function (error) {
    console.log("clientWebSocket.onclose", clientWebSocket, error);
    events("聊天会话关闭！");
  }
  function events(responseEvent) {
    document.querySelector(".chat").innerHTML += responseEvent + "<br>";
  }
</script>
</body>
</html>
```

9.7.6 运行程序

先运行服务器端主程序，再运行客户端 WSClient，正常运行后控制台的主要输出结果如图 9-14 所示。程序正常运行时，在浏览器中输入 localhost:8080/websocket-client.html，结果如图 9-15 所示。关闭程序后浏览器中的输出结果如图 9-16 所示。

服务端返回：小明，　你好

图 9-14　控制台中主要输出结果

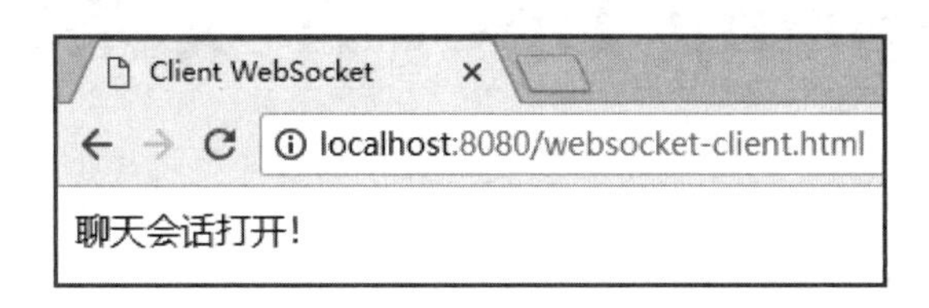

图 9-15　在浏览器中输入 localhost:8080/websocket-client.html 后的结果

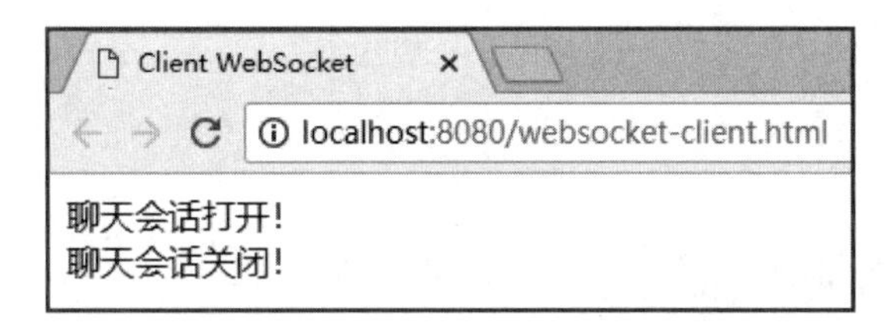

图 9-16　关闭程序后浏览器中的输出结果

习题 9

简答题

1．简述对 WebFlux 的理解。

2．简述对 WebFlux 编程模型的理解。

3．简述对 Redis 的理解。

4．简述对 WebSocket 的理解。

实验题

1．实现 WebFlux 的简单应用。

2．实现基于 WebFlux 的 RESTful 服务。

3．实现基于 WebFlux 访问 MongoDB 数据库。

4．实现基于 WebFlux 使用 Thymeleaf。

5．实现基于 WebFlux 访问 Redis 数据库。

6．实现基于 WebFlux 使用 WebSocket。

第 10 章

Spring Boot 开发案例

本章结合一个案例说明 Spring Boot 的开发过程。

10.1 案例分析

10.1.1 主要界面

用户登录界面如图 10-1 所示；用户注册界面如图 10-2 所示；用户登录后的主界面如图 10-3 所示；增加新的课程类型界面如图 10-4 所示；管理课程类型界面如图 10-5 所示；增加新的课程界面如图 10-6 所示；管理课程界面如图 10-7 所示；用户退出系统前的提示界面如图 10-8 所示。

Spring Boot
课程管理系统用户登录
用户姓名：请输入姓名
登录密码：请输入密码
登 录 注册
© 2018 参考网上实例
当前时间为 2018年10月9日 17:40 星期二

图 10-1 用户登录界面

图 10-2 用户注册界面

图 10-3 用户登录后的主界面

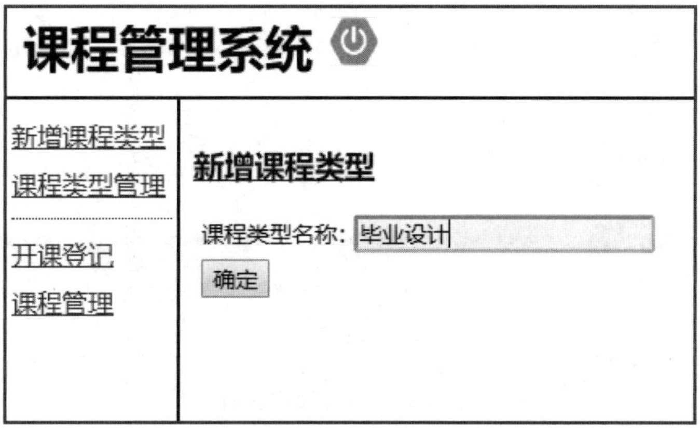

图 10-4 增加新的课程类型界面

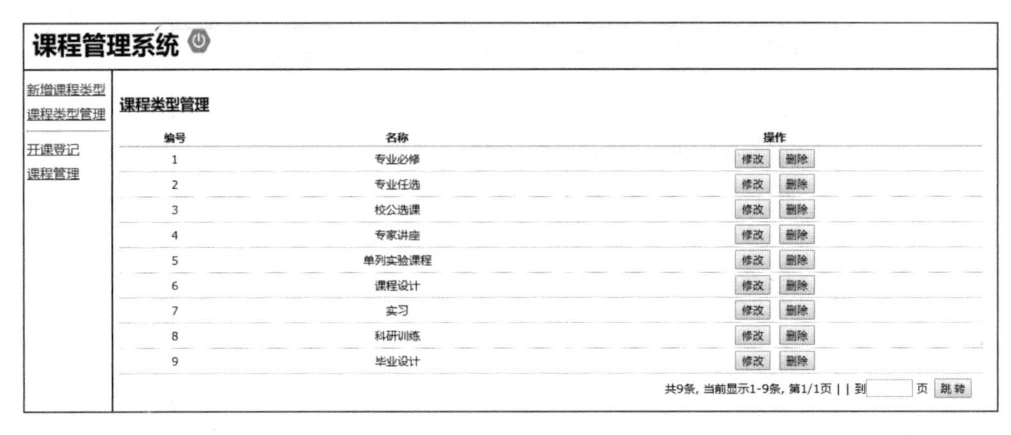

图 10-5　管理课程类型界面

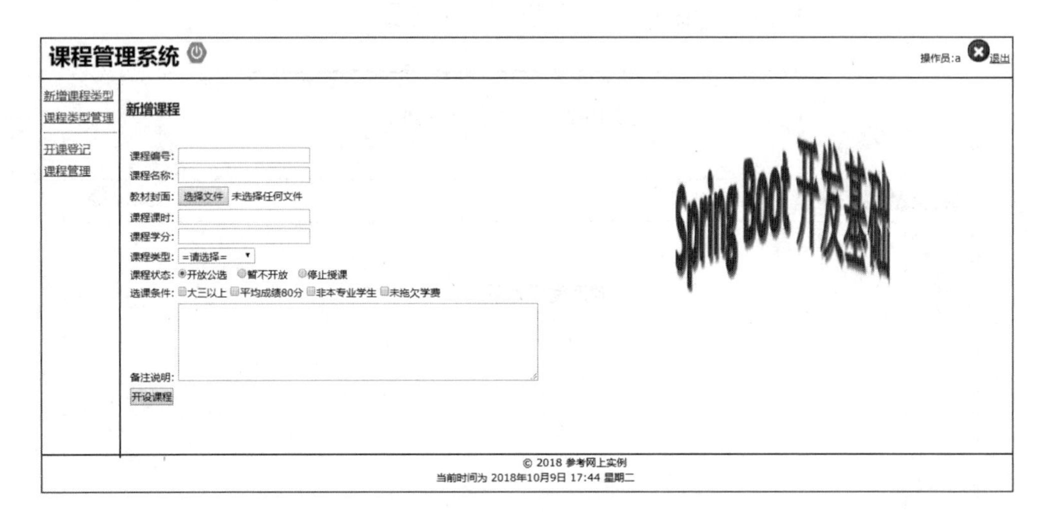

图 10-6　增加新的课程界面

课程管理系统　操作员:a　退出

新增课程类型
课程类型管理
开课登记
课程管理

课程管理

课程名称:　学分范围: -　课程类型: =请选择=　查询

序号	编号	名称	课时	学分	类型	状态	选课要求	备注	操作
1	1	SSH开发基础	54	2.0	专业必修	开放公选	大三以上	Struts+Spring+Hibernate	修改 删除
2	2	SpringMVC	54	3.0	专业必修	开放公选	大三以上	SpringMVC	修改 删除
3	3	SpringBoot开发基础	54	2.0	专业必修	开放公选	大三以上	Spring Boot	修改 删除
4	4	SSM开发基础	54	2.0	专业必修	开放公选	大三以上	Spring+SpringMVC+MyBatis	修改 删除

共5条，当前显示1-4条，第1/2页 | 下一页 末页 | 到 页 跳转

© 2018 参考网上实例
当前时间为 2018年10月9日 17:44 星期二

图 10-7　管理课程界面

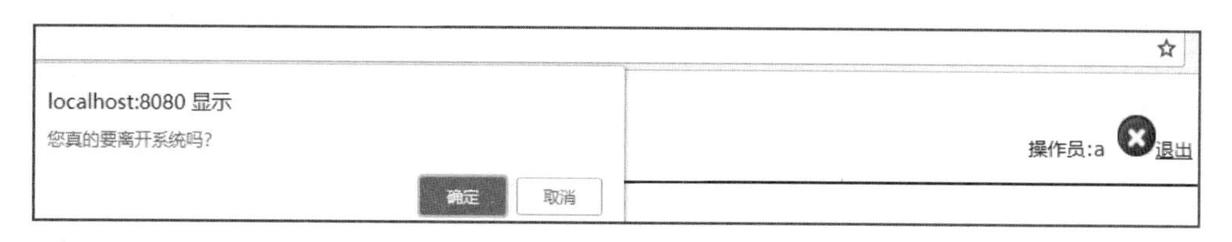

图 10-8　用户退出系统前的提示界面

10.1.2　主要功能与数据库介绍

系统的主要功能包括用户登录、用户注册、增加课程类型、管理课程类型（修改和删除课程类型）、增加课程、管理课程（修改和删除课程）、用户退出系统等功能。为了更好地实现系统界面，需要用到页面片段的定义和使用功能。为了更好地显示课程类型和课程，需要用到分页功能。

为了存储数据，需要用到数据库。本案例中用的数据库是 MySQL 数据库。用到的主要的表有用户表 tbl_users、课程表 tbl_course、课程类型表 tbl_course_type。创建这些表和向表中插入记录的 SQL 语句如例 10-1 所示。

【例 10-1】 创建表和向表中插入记录 SQL 语句的代码示例。

```
create table tbl_course(
  course_no varchar(50) primary key,
  course_name varchar(100) not null,
  course_hours int not null,
  type_id int not null,
  course_status varchar(1) not null,
  course_reqs varchar(20) not null,
  course_point decimal(3,1),
  course_memo varchar(1000),
  course_textbook_pic mediumblob,
  constraint FK_COURSE_TYPE FOREIGN KEY (type_id) references tbl_course_
  type(type_id)
);
create table tbl_course_type(
  type_id int primary key auto_increment,
  type_name varchar(30) not null
 );
insert into tbl_course_type(type_name) values('专业必修');
insert into tbl_course_type(type_name) values('专业任选');
insert into tbl_course_type(type_name) values('校选课');
insert into tbl_course_type(type_name) values('专家讲座');
create table tbl_users(
  user_no   varchar(20) primary key,
  user_pwd  varchar(1000) not null,
  user_name  varchar(100) not null
```

```
);
insert into tbl_users values('000101','123456','王涛');
insert into tbl_users values('000102','123456','张三');
alter table tbl_course add course_textbook_pic mediumblob;
```

10.2 案例实现

视频讲解

10.2.1 添加依赖

在 pom.xml 文件中<dependencies>和</dependencies>之间添加依赖，代码如例 10-2 所示。

【例 10-2】 添加依赖的代码示例。

```
<dependency>
        <groupId>org.springframework.boot</groupId>
        <artifactId>spring-boot-starter</artifactId>
</dependency>
<dependency>
        <groupId>org.springframework.boot</groupId>
        <artifactId>spring-boot-starter-test</artifactId>
        <scope>test</scope>
</dependency>
<dependency>
        <groupId>org.springframework.boot</groupId>
        <artifactId>spring-boot-starter-aop</artifactId>
        <exclusions>
            <exclusion>
                <groupId>org.springframework.boot</groupId>
                <artifactId>spring-boot-starter-logging</artifactId>
            </exclusion>
        </exclusions>
</dependency>
<dependency>
        <groupId>org.slf4j</groupId>
        <artifactId>slf4j-api</artifactId>
</dependency>
<dependency>
        <groupId>org.springframework.boot</groupId>
        <artifactId>spring-boot-starter-web</artifactId>
</dependency>
<dependency>
        <groupId>org.springframework.boot</groupId>
        <artifactId>spring-boot-devtools</artifactId>
</dependency>
```

```
<dependency>
        <groupId>org.springframework.boot</groupId>
        <artifactId>spring-boot-starter-thymeleaf</artifactId>
</dependency>
<dependency>
        <groupId>javax.persistence</groupId>
        <artifactId>persistence-api</artifactId>
        <version>1.0</version>
</dependency>
<dependency>
        <groupId>org.springframework</groupId>
        <artifactId>spring-context-support</artifactId>
</dependency>
<dependency>
        <groupId>org.springframework</groupId>
        <artifactId>spring-tx</artifactId>
</dependency>
<dependency>
        <groupId>tk.mybatis</groupId>
        <artifactId>mapper-spring-boot-starter</artifactId>
        <version>2.0.4</version>
</dependency>
<dependency>
        <groupId>com.github.pagehelper</groupId>
        <artifactId>pagehelper-spring-boot-starter</artifactId>
        <version>1.2.6</version>
</dependency>
<dependency>
        <groupId>org.apache.commons</groupId>
        <artifactId>commons-lang3</artifactId>
</dependency>
<dependency>
        <groupId>mysql</groupId>
        <artifactId>mysql-connector-java</artifactId>
</dependency>
<dependency>
        <groupId>com.alibaba</groupId>
        <artifactId>druid-spring-boot-starter</artifactId>
</dependency>
```

10.2.2　创建类 User、CourseType 和 Course

实体类 User 的代码如例 10-3 所示。

【例 10-3】 实体类 User 的代码示例。

```
package xiao.ze.demo.entity;
import org.springframework.stereotype.Component;
import javax.persistence.*;
import javax.validation.constraints.NotEmpty;
import java.io.Serializable;
@Component
@Table(name="tbl_users")
public class User implements Serializable{
    @Id
    @GeneratedValue(strategy = GenerationType.IDENTITY)
    @Column(name="user_no")
    private int userNo;
    @Column(name="user_name")
    private String userName;
    private String userPwd;
    public int getUserNo() {
        return userNo;
    }
    public void setUserNo(int userNo) {
        this.userNo = userNo;
    }
    public String getUserName() {
        return userName;
    }
    public void setUserName(String userName) {
        this.userName = userName;
    }
    public String getUserPwd() {
        return userPwd;
    }
    public void setUserPwd(String userPwd) {
        this.userPwd = userPwd;
    }
}
```

实体类 CourseType 的代码如例 10-4 所示。

【例 10-4】 实体类 CourseType 的代码示例。

```
package xiao.ze.demo.entity;
import org.springframework.stereotype.Component;
import javax.persistence.GeneratedValue;
import javax.persistence.GenerationType;
import javax.persistence.Id;
import javax.persistence.Table;
import java.io.Serializable;
```

```
@Component
@Table(name="tbl_course_type")
public class CourseType implements Serializable{
    @Id
    @GeneratedValue(strategy = GenerationType.IDENTITY)
    private Integer typeId;
    private String  typeName;
    public Integer getTypeId() {
        return typeId;
    }
    public void setTypeId(Integer typeId) {
        this.typeId = typeId;
    }
    public String getTypeName() {
        return typeName;
    }
    public void setTypeName(String typeName) {
        this.typeName = typeName;
    }
    @Override
    public String toString() {
        return "CourseType [typeId=" + typeId + ", typeName=" + typeName + "]";
    }
}
```

实体类 Course 的代码如例 10-5 所示。

【例 10-5】 实体类 Course 的代码示例。

```
package xiao.ze.demo.entity;
import org.springframework.stereotype.Component;
import java.io.Serializable;
@Component
public class Course implements Serializable{
    private String courseNo;
    private String courseName;
    private Integer courseHours;
    private String courseStatus;
    private Double coursePoint;
    private String[] courseReqs;
    private String reqs;
    private String courseMemo;
    private byte[] courseTextbookPic;   //教材封面
    private CourseType courseType;
    public String getCourseNo() {
        return courseNo;
    }
    public void setCourseNo(String courseNo) {
```

```
        this.courseNo = courseNo;
    }
    public String getCourseName() {
        return courseName;
    }
    public void setCourseName(String courseName) {
        this.courseName = courseName;
    }
    public Integer getCourseHours() {
        return courseHours;
    }
    public void setCourseHours(Integer courseHours) {
        this.courseHours = courseHours;
    }
    public String[] getCourseReqs() {
        return courseReqs;
    }
    public void setCourseReqs(String[] courseReqs) {
        this.courseReqs = courseReqs;
        StringBuffer sb = new StringBuffer();
        for(String req:courseReqs){
            sb.append(req).append("|");
        }
        sb.deleteCharAt(sb.length()-1);
        this.reqs = sb.toString();
    }
    public CourseType getCourseType() {
        return courseType;
    }
    public void setCourseType(CourseType courseType) {
        this.courseType = courseType;
    }
    public String getCourseStatus() {
        return courseStatus;
    }
    public void setCourseStatus(String courseStatus) {
        this.courseStatus = courseStatus;
    }
    public Double getCoursePoint() {
        return coursePoint;
    }
    public void setCoursePoint(Double coursePoint) {
        this.coursePoint = coursePoint;
    }
    public String getCourseMemo() {
        return courseMemo;
```

```
    }
    public void setCourseMemo(String courseMemo) {
        this.courseMemo = courseMemo;
    }
    public byte[] getCourseTextbookPic() {
        return courseTextbookPic;
    }
    public void setCourseTextbookPic(byte[] courseTextbookPic) {
        this.courseTextbookPic = courseTextbookPic;
    }
    public String getReqs() {
        return reqs;
    }
    public void setReqs(String reqs) {
        this.reqs = reqs;
        this.courseReqs = this.reqs.split("\\|");
    }
}
```

10.2.3　创建 Service 接口

接口 UserService 的代码如例 10-6 所示。

【例 10-6】 接口 UserService 的代码示例。

```
package xiao.ze.demo.service;
import xiao.ze.demo.entity.User;
import java.util.List;
public interface UserService {
    List<User> loadUserByUserName(String userName);
    void addUser(User user);
}
```

接口 CourseTypeService 的代码如例 10-7 所示。

【例 10-7】 接口 CourseTypeService 的代码示例。

```
package xiao.ze.demo.service;
import xiao.ze.demo.entity.CourseType;
import java.util.List;
public interface CourseTypeService {
    void addCourseType(CourseType courseType);
    void removeCourseType(Integer typeId);
    void updateCourseType(CourseType courseType);
    CourseType getCourseTypeById(Integer typeId);
    List<CourseType> loadAll();
}
```

接口 CourseService 的代码如例 10-8 所示。

【例 10-8】 接口 CourseService 的代码示例。

```
package xiao.ze.demo.service;
import xiao.ze.demo.entity.Course;
import xiao.ze.demo.utils.CourseQueryHelper;
import java.util.List;
public interface CourseService {
    void addCourse(Course course);
    boolean removeCourseByNo(String courseNo);
    void updateCourse(Course course);
    Course loadCourseByNo(String courseNo);
    List<Course> loadScopedCourses(CourseQueryHelper helper);
    byte[] getTextbookPic(String courseNo);
}
```

10.2.4 创建 Service 接口实现类

接口 UserService 实现类 UserServiceImpl 的代码如例 10-9 所示。

【例 10-9】 接口 UserService 实现类 UserServiceImpl 的代码示例。

```
package xiao.ze.demo.service.impl;
import org.springframework.stereotype.Service;
import org.springframework.transaction.annotation.Transactional;
import xiao.ze.demo.entity.User;
import xiao.ze.demo.mapper.UserMapper;
import xiao.ze.demo.service.UserService;
import javax.annotation.Resource;
import java.util.List;
@Service
@Transactional(rollbackFor = Exception.class)
public class UserServiceImpl implements UserService {
    @Resource
    private UserMapper userMapper ;
    @Override
    public  List<User>  loadUserByUserName(String userName){
        List<User> users = null;
        User user = userMapper.loadUserByUserName(userName);
        users =userMapper.select(user);
        return users;
    }
    @Override
    public void addUser(User user) {
        userMapper.insert(user);
    }
}
```

接口 CourseTypeService 实现类 CourseTypeServiceImpl 的代码如例 10-10 所示。

【例 10-10】 接口 CourseTypeService 实现类 CourseTypeServiceImpl 的代码示例。

```
package xiao.ze.demo.service.impl;
import org.springframework.stereotype.Service;
import org.springframework.transaction.annotation.Transactional;
import xiao.ze.demo.entity.CourseType;
import xiao.ze.demo.mapper.CourseMapper;
import xiao.ze.demo.mapper.CourseTypeMapper;
import xiao.ze.demo.service.CourseTypeService;
import javax.annotation.Resource;
import java.util.List;
@Service
@Transactional(rollbackFor = Exception.class)
public class CourseTypeServiceImpl implements CourseTypeService {
    @Resource
    private CourseTypeMapper courseTypeMapper;
    @Resource
    private CourseMapper courseMapper;
    @Override
    public void addCourseType(CourseType courseType) {
        courseTypeMapper.insert(courseType);
    }
    @Override
    public void removeCourseType(Integer typeId) {
        if(courseMapper.loadCourseByTypeId(typeId)!=null){
            courseMapper.removeCourseByTypeId(typeId);
        }
        courseTypeMapper.deleteByPrimaryKey(typeId);
    }
    @Override
    public void updateCourseType(CourseType courseType) {
        courseTypeMapper.updateByPrimaryKey(courseType);
    }
    @Override
    public CourseType getCourseTypeById(Integer typeId) {
        return courseTypeMapper.selectByPrimaryKey(typeId);
    }
    @Override
    public List<CourseType> loadAll() {
        return courseTypeMapper.selectAll();
    }
}
```

接口 CourseService 实现类 CourseServiceImpl 的代码如例 10-11 所示。

【例 10-11】 接口 CourseService 实现类 CourseServiceImpl 的代码示例。

```
package xiao.ze.demo.service.impl;
import org.springframework.stereotype.Service;
import org.springframework.transaction.annotation.Transactional;
import xiao.ze.demo.entity.Course;
import xiao.ze.demo.mapper.CourseMapper;
import xiao.ze.demo.service.CourseService;
import xiao.ze.demo.utils.CourseQueryHelper;
import javax.annotation.Resource;
import java.util.HashMap;
import java.util.List;
import java.util.Map;
@Service
@Transactional(rollbackFor = Exception.class)
public class CourseServiceImpl implements CourseService {
    @Resource
    private CourseMapper courseMapper;
    @Override
    public void addCourse(Course course) {
        courseMapper.addCourse(course);
    }
    @Override
    public boolean removeCourseByNo(String courseNo) {
        courseMapper.removeCourseByNo(courseNo) ;
        return true;
    }
    @Override
    public void updateCourse(Course course) {
        String[] courseReq = course.getCourseReqs();
        if (courseReq != null && courseReq.length > 0) {
            courseMapper.updateCourse(course);
        } else {
            course.setReqs("");
            courseMapper.updateCourse(course);
        }
    }
    @Override
    public Course loadCourseByNo(String courseNo) {
        Course course=new Course();
        course=null;
        if(courseNo!=null) {
            course = courseMapper.loadCourseByNo(courseNo);
        }
        return course;
    }
    @Override
```

```
    public List<Course> loadScopedCourses(CourseQueryHelper helper) {
        Map<String,Object> map = new HashMap<>(16);
        map=getQueryHelper(helper);
        List<Course> list = courseMapper.loadScopedCourses(map);
        return list;
    }
    @Override
    public byte[] getTextbookPic(String courseNo) {
        byte[] textBookPic = null;
        Course course = courseMapper.loadCourseByNo(courseNo);
        textBookPic = course.getCourseTextbookPic();
        return textBookPic;
    }
    private Map<String,Object> getQueryHelper(CourseQueryHelper helper) {
        Map<String,Object> map = new HashMap<>(16);
        if(helper.getQryCourseName()!=null){
            map.put("qryCourseName", helper.getQryCourseName());
        }
        if(helper.getQryEndPoint()!=null){
            map.put("qryEndPoint", helper.getQryEndPoint());
        }
        if(helper.getQryStartPoint()!=null){
            map.put("qryStartPoint", helper.getQryStartPoint());
        }
        if((helper.getQryCourseType()!=null)&&(!"".equals(helper.
        getQryCourseType()))){
            map.put("typeId", Integer.parseInt(helper.getQryCourseType()));
        }
        return map;
    }
}
```

10.2.5　创建 Mapper 接口

接口 UserMapper 的代码如例 10-12 所示。

【例 10-12】　接口 UserMapper 的代码示例。

```
package xiao.ze.demo.mapper;
import tk.mybatis.mapper.common.Mapper;
import xiao.ze.demo.entity.User;
interface UserMapper extends Mapper<User> {
    User loadUserByUserName(String userName);
}
```

接口 CourseTypeMapper 的代码如例 10-13 所示。

【例 10-13】 接口 CourseTypeMapper 的代码示例。

```
package xiao.ze.demo.mapper;
import tk.mybatis.mapper.common.Mapper;
import xiao.ze.demo.entity.CourseType;
public interface CourseTypeMapper extends Mapper<CourseType> {
}
```

接口 CourseMapper 的代码如例 10-14 所示。

【例 10-14】 接口 CourseMapper 的代码示例。

```
package xiao.ze.demo.mapper;
import xiao.ze.demo.entity.Course;
import java.util.List;
import java.util.Map;
public interface CourseMapper {
    void addCourse(Course course);
    boolean removeCourseByNo(String courseNo);
    boolean removeCourseByTypeId(Integer typeId);
    void updateCourse(Course course);
    Course loadCourseByNo(String courseNo);
    List<String> loadCourseByTypeId(Integer typeId);
    List<Course> loadScopedCourses(Map map);
}
```

10.2.6 创建类 WebLogAspect

日志类 WebLogAspect 的代码如例 10-15 所示。

【例 10-15】 日志类 WebLogAspect 的代码示例。

```
import org.aspectj.lang.annotation.Aspect;
import org.slf4j.Logger;
import org.slf4j.LoggerFactory;
import org.springframework.stereotype.Component;
import java.util.Arrays;
import java.util.Enumeration;
import javax.servlet.http.HttpServletRequest;
import org.aspectj.lang.JoinPoint;
import org.aspectj.lang.annotation.AfterReturning;
import org.aspectj.lang.annotation.Before;
import org.aspectj.lang.annotation.Pointcut;
import org.springframework.web.context.request.RequestContextHolder;
import org.springframework.web.context.request.ServletRequestAttributes;
//实现 Web 层的日志切面
@Aspect
@Component
```

```
public class WebLogAspect {
    private Logger logger =  LoggerFactory.getLogger(this.getClass());
    ThreadLocal<Long> startTime = new ThreadLocal<Long>();
    @Pointcut("execution(* xiao.ze.demo.service.impl.*.*(..))")
    public void webLog(){}
    @Before("webLog()")
    public void doBefore(JoinPoint joinPoint){
        startTime.set(System.currentTimeMillis());
        //接收到请求，记录请求内容
        logger.info("WebLogAspect.doBefore()");
        ServletRequestAttributes attributes = (ServletRequestAttributes)
        RequestContextHolder.getRequestAttributes();
        HttpServletRequest request = attributes.getRequest();
        //记录请求内容
        logger.info("URL : " + request.getRequestURL().toString());
        logger.info("HTTP_METHOD : " + request.getMethod());
        logger.info("IP : " + request.getRemoteAddr());
        logger.info("CLASS_METHOD : " + joinPoint.getSignature().
        getDeclaringTypeName() + "." + joinPoint.getSignature().getName());
        logger.info("ARGS : " + Arrays.toString(joinPoint.getArgs()));
        //获取所有参数方法
        Enumeration<String> enu=request.getParameterNames();
        while(enu.hasMoreElements()){
            String paraName=(String)enu.nextElement();
            System.out.println(paraName+": "+request.getParameter(paraName));
        }
    }
    @AfterReturning("webLog()")
    public void  doAfterReturning(JoinPoint joinPoint){
        //处理完请求，返回内容
        logger.info("WebLogAspect.doAfterReturning()");
        logger.info("耗时（毫秒） : " + (System.currentTimeMillis() -
        startTime.get()));
    }
}
```

10.2.7　创建类 CourseQueryHelper

辅助分页类 CourseQueryHelper 的代码如例 10-16 所示。

【例 10-16】 辅助分页类 CourseQueryHelper 的代码示例。

```
package xiao.ze.demo.utils;
public class CourseQueryHelper {
    private String qryCourseName;
    private Double qryStartPoint;
```

```
    private Double qryEndPoint;
    private String qryCourseType;
    public String getQryCourseName() {
        return qryCourseName;
    }
    public void setQryCourseName(String qryCourseName) {
        this.qryCourseName = qryCourseName;
    }
    public Double getQryStartPoint() {
        return qryStartPoint;
    }
    public void setQryStartPoint(Double qryStartPoint) {
        this.qryStartPoint = qryStartPoint;
    }
    public Double getQryEndPoint() {
        return qryEndPoint;
    }
    public void setQryEndPoint(Double qryEndPoint) {
        this.qryEndPoint = qryEndPoint;
    }
    public String getQryCourseType() {
        return qryCourseType;
    }
    public void setQryCourseType(String qryCourseType) {
        this.qryCourseType = qryCourseType;
    }
}
```

10.2.8 创建控制器类

控制器类 IndexController 的代码如例 10-17 所示。

【例 10-17】 控制器类 IndexController 的代码示例。

```
package xiao.ze.demo.controller;
import org.springframework.stereotype.Controller;
import org.springframework.web.bind.annotation.GetMapping;
@Controller
public class IndexController {
    @GetMapping("/")
    public String root() {
        return "index";
    }
}
```

控制器类 SecurityController 的代码如例 10-18 所示。

【例 10-18】　控制器类 SecurityController 的代码示例。

```
package xiao.ze.demo.controller;
import java.util.List;
import java.util.Map;
import org.springframework.beans.factory.annotation.Autowired;
import org.springframework.stereotype.Controller;
import org.springframework.web.bind.annotation.*;
import xiao.ze.demo.entity.User;
import xiao.ze.demo.service.UserService;
import javax.servlet.http.HttpServletRequest;
@Controller
@RequestMapping("/security")
public class SecurityController {
    @Autowired
    private UserService userService ;
    @RequestMapping("/index")
    public String root() {
        return "index";
    }
    @GetMapping("/toLogin")
    public String toLogin(Map<String, Object> map) {
        map.put("user", new User());
        return "login";
    }
    //注册页面
    @RequestMapping("/register")
    public String register(){
        return "register";
    }
    //注册协议页面
    @RequestMapping("/readdoc")
    public String readdoc(){
        return "readdoc";
    }
    //注册方法
    @RequestMapping("/addregister")
    public String register(HttpServletRequest request){
        String username = request.getParameter("username");
        String password = request.getParameter("password");
        String password2 = request.getParameter("password2");
        if (password.equals(password2)){
            User userEntity = new User();
            userEntity.setUserName(username);
            userEntity.setUserPwd(password);
```

```
            userService.addUser(userEntity);
            return "login";
        }else {
            return "register";
        }
    }
    @PostMapping(value="/login")
    public String login(  User user,Map<String, Object> map ) {
        if(userService.loadUserByUserName(user.getUserName())!=null){
          List<User>  lus=userService.loadUserByUserName(user.
          getUserName());
            if(lus.get(0).getUserPwd().equals(user.getUserPwd())){
                map.put("user",lus.get(0));
                return "main";
            }
        }
        return "login";
    }
    @GetMapping("/mainController")
    public String main(){
        return "main";
    }
    @GetMapping("/logout")
    public String logout(){
        return "redirect:/security/toLogin";
    }
}
```

控制器类 CourseTypeController 的代码如例 10-19 所示。

【例 10-19】 控制器类 CourseTypeController 的代码示例。

```
package xiao.ze.demo.controller;
import com.github.pagehelper.PageHelper;
import com.github.pagehelper.PageInfo;
import org.springframework.beans.factory.annotation.Autowired;
import org.springframework.stereotype.Controller;
import org.springframework.web.bind.annotation.*;
import xiao.ze.demo.entity.CourseType;
import xiao.ze.demo.service.CourseTypeService;
import java.util.List;
import java.util.Map;
@Controller
@RequestMapping("/courseType")
public class CourseTypeController {
    @Autowired
    private CourseTypeService courseTypeService ;
```

```
    @GetMapping("/toInput")
    public String input(Map<String, Object> map) {
        map.put("courseType", new CourseType());
        return "courseType/input_course_type";
    }
    @PostMapping(value="/create")
    public String create(CourseType courseType) {
        courseTypeService.addCourseType(courseType);
        return "redirect:/courseType/list";
    }
    @GetMapping("/list")
    public String list(Map<String, Object> map, @RequestParam(value=
    "pageNo", required=false, defaultValue="1") String pageNoStr) {
        int pageNo = 1;
        pageNo = Integer.parseInt(pageNoStr);
        if(pageNo < 1){
            pageNo = 1;
        }
        PageHelper.startPage(pageNo, 10);
        List<CourseType> courseTypeList = courseTypeService.loadAll();
        PageInfo<CourseType> page=new PageInfo<CourseType>(courseTypeList);
        map.put("page", page);
        return "courseType/list_course_type";
    }
    @DeleteMapping(value="/remove/{typeId}")
    public String remove(@PathVariable("typeId") Integer typeId) {
        courseTypeService.removeCourseType(typeId);
        return "redirect:/courseType/list";
    }
    @GetMapping(value="/preUpdate/{typeId}")
    public String preUpdate(@PathVariable("typeId") Integer typeId,
    Map<String, Object> map) {
        System.out.println(courseTypeService.getCourseTypeById(typeId));
        map.put("courseType", courseTypeService.getCourseTypeById(typeId));
        return "courseType/update_course_type";
    }
    @PutMapping(value="/update")
    public String update(CourseType courseType) {
        courseTypeService.updateCourseType(courseType);
        return "redirect:/courseType/list";
    }
}
```

控制器类 CourseController 的代码如例 10-20 所示。

【例 10-20】 控制器类 CourseController 的代码示例。

```
package xiao.ze.demo.controller;
import com.github.pagehelper.PageHelper;
import com.github.pagehelper.PageInfo;
import org.springframework.beans.factory.annotation.Autowired;
import org.springframework.stereotype.Controller;
import org.springframework.web.bind.annotation.*;
import org.springframework.web.multipart.MultipartFile;
import xiao.ze.demo.entity.Course;
import xiao.ze.demo.service.CourseService;
import xiao.ze.demo.service.CourseTypeService;
import xiao.ze.demo.utils.CourseQueryHelper;
import javax.servlet.ServletOutputStream;
import javax.servlet.http.HttpServletRequest;
import javax.servlet.http.HttpServletResponse;
import java.io.File;
import java.io.FileInputStream;
import java.util.List;
import java.util.Map;
@Controller
@RequestMapping("/course")
public class CourseController {
    @Autowired
    private CourseService courseService;
    @Autowired
    private CourseTypeService courseTypeService ;
    @ModelAttribute
    public void getCourse(@RequestParam(value="courseNo",required=false)
    String courseNo, Map<String, Object> map,Course course){
        course=courseService.loadCourseByNo(courseNo);
        if(courseNo != null&&course!= null){
           map.put("course", course);
        }
    }
    @GetMapping("/toInput")
    public String toInput(Map<String, Object> map,Course course) {
        map.put("courseTypeList", courseTypeService.loadAll());
        course.setCourseStatus("O");
        course.setCourseReqs(new String[]{"a","b"});
        map.put("course", course);
        return "course/input_course";
    }
    @PostMapping(value="/create")
    public String create(@RequestParam("coursetextbookpic") MultipartFile
    file, Course course, Map<String, Object> map) throws Exception{
```

```
    //读取文件数据，转成字节数组
    if(file!=null){
        course.setCourseTextbookPic(file.getBytes());
    }
    try{
        courseService.addCourse(course);
        System.out.println("你好");
    }catch(Exception e){
        map.put("exceptionMessage", e.getMessage());
        map.put("courseTypeList", courseTypeService.loadAll());
        return "course/input_course";
    }
    return "redirect:/course/list";
}
@RequestMapping("/list")
public String list(@RequestParam(value="pageNo", required=false,
defaultValue="1") String pageNoStr,
    Map<String, Object> map, CourseQueryHelper helper) {
    int pageNo = 1;
    //对 pageNo 的校验
    pageNo = Integer.parseInt(pageNoStr);
    if(pageNo < 1){
        pageNo = 1;
    }
    PageHelper.startPage(pageNo, 4);
    List<Course> courselist = courseService.loadScopedCourses(helper);
    PageInfo<Course> page=new PageInfo<Course>(courselist);
    map.put("courseTypeList", courseTypeService.loadAll());
    map.put("page", page);
    map.put("helper", helper);
    return "course/list_course";
}
@DeleteMapping(value="/remove/{courseNo}")
public String remove(@PathVariable("courseNo") String courseNo) {
    courseService.removeCourseByNo(courseNo);
    return "redirect:/course/list";
}
@GetMapping(value="/preUpdate/{courseNo}")
public  String  preUpdate(@PathVariable("courseNo")  String  courseNo,
Map<String, Object> map) {
    map.put("course" ,courseService.loadCourseByNo(courseNo));
    map.put("courseTypeList", courseTypeService.loadAll());
    return "course/update_course";
}
@PostMapping(value="/update")
```

```
    public String update(@RequestParam("coursetextbookpic") MultipartFile
    file, Course course, Map<String, Object> map) throws Exception{
        //读取多段提交的文件数据，转成字节数组
        if(file.getBytes().length>0){
            course.setCourseTextbookPic(file.getBytes());
        }
        try{
            courseService.updateCourse(course);
        }catch(Exception e){
            map.put("exceptionMessage", e.getMessage());
            map.put("courseTypeList", courseTypeService.loadAll());
            return "/course/update_course";
        }
        return "redirect:/course/list";
    }
    @GetMapping("/getPic/{courseNo}")
    public String getPic(@PathVariable("courseNo") String courseNo,
    HttpServletRequest request, HttpServletResponse response) throws
    Exception{
        byte[] textBookPic = courseService.getTextbookPic(courseNo);
        if(textBookPic==null){
            String path = request.getSession().getServletContext().
            getRealPath("/pics/default.jpg");
            FileInputStream fis = new FileInputStream(new File(path));
            textBookPic = new byte[fis.available()];
            fis.read(textBookPic);
        }
        //向浏览器发通知，我要发送的是图片
        response.setContentType("image/jpeg");
        ServletOutputStream sos=response.getOutputStream();
        sos.write(textBookPic);
        sos.flush();
        sos.close();
        return null;
    }
}
```

10.2.9 修改入口类

修改后的入口类代码如例 10-21 所示。

【例 10-21】 修改后的入口类的代码示例。

```
package xiao.ze.demo.start;
import tk.mybatis.spring.annotation.MapperScan;
import org.springframework.boot.SpringApplication;
```

```
import org.springframework.boot.autoconfigure.SpringBootApplication;
import org.springframework.context.annotation.ComponentScan;
import org.springframework.transaction.annotation.EnableTransactionManagement;
@ComponentScan(basePackages = "xiao.ze.demo")
@SpringBootApplication
@MapperScan(basePackages ="xiao.ze.demo.mapper")
@EnableTransactionManagement
public class App {
    public static void main(String[] args) {
        SpringApplication.run(App.class, args);
    }
}
```

10.2.10　创建 XML 文件

文件 CourseMapper.xml 的代码如例 10-22 所示。

【例 10-22】 文件 CourseMapper.xml 的代码示例。

```
<?xml version="1.0" encoding="UTF-8" ?>
<!DOCTYPE mapper
        PUBLIC "-//mybatis.org//DTD Mapper 3.0//EN"
        "http://mybatis.org/dtd/mybatis-3-mapper.dtd">
<!--配置命名空间，区别名称-->
<mapper namespace="xiao.ze.demo.mapper.CourseMapper">
    <!--SQL 片段-->
    <sql id="cols">course_no,
        course_name,
        course_hours,
        type_id,
        course_status,
        course_reqs,
        course_point,
        course_memo,
        course_textbook_pic
    </sql>
    <!--中间，对象的属性和结果集的字段之间的对应关系-->
    <resultMap type="xiao.ze.demo.entity.Course" id="courseRM">
        <!--主键映射-->
        <id property="courseNo" column="course_no"/>
        <!--普通字段 property 指实体的属性；column 结果集的字段名称-->
        <result property="courseName" column="course_name"/>
        <result property="courseHours" column="course_hours"/>
        <result property="courseStatus" column="course_status"/>
        <result property="reqs" column="course_reqs"/>
        <result property="coursePoint" column="course_point"/>
```

```
        <result property="courseMemo" column="course_memo"/>
        <result property="courseTextbookPic" column="course_textbook_pic"/>
        <!--对象关联-->
        <association property="courseType" javaType="xiao.ze.demo.entity.
        CourseType">
            <!--主键映射-->
            <id property="typeId" column="type_id"/>
            <!--普通字段 property 指实体的属性，column 结果集的字段名称-->
            <result property="typeName" column="type_name"/>
        </association>
    </resultMap>
    <!--新增-->
    <insert id="addCourse" parameterType="xiao.ze.demo.entity.Course">
        insert into tbl_course
        (<include refid="cols"/>)
        values
        (#{courseNo},#{courseName},#{courseHours},
        #{courseType.typeId},#{courseStatus},#{reqs},
        #{coursePoint},#{courseMemo},#{courseTextbookPic,jdbcType=BLOB})
    </insert>
    <delete id="removeCourseByNo" parameterType="string">
        delete from tbl_course
        where course_no = #{courseNo}
    </delete>
    <delete id="removeCourseByTypeId" parameterType="int">
        delete from tbl_course
        where type_id = #{typeId}
    </delete>
    <!--修改-->
    <update id="updateCourse" parameterType="xiao.ze.demo.entity.Course" >
        update tbl_course
        <set>
            <if test="courseName!=null">course_name=#{courseName},</if>
            <if test="courseHours!=null">course_hours=#{courseHours},</if>
            <if test="courseType!=null">type_id=#{courseType.typeId},</if>
            <if test="courseStatus!=null">course_status=#{courseStatus},</if>
            <if test="reqs!=null">course_reqs=#{reqs},</if>
            <if test="coursePoint!=null">course_point=#{coursePoint},</if>
            <if test="courseMemo!=null">course_memo=#{courseMemo},</if>
            <if test="courseTextbookPic!=null">course_textbook_pic=#
            {courseTextbookPic,jdbcType=BLOB},</if>
        </set>
        where course_no = #{courseNo}
    </update>
    <select id="loadCourseByNo" parameterType="string" resultType="xiao.
    ze.demo.entity.Course" resultMap="courseRM">
```

```
        select
        p.course_no,p.course_name,p.course_hours,p.course_status,
        p.course_reqs,p.course_point,p.course_memo,p.course_textbook_pic,
        b.type_id,b.type_name
        from tbl_course p
        left join tbl_course_type b
        on p.type_id=b.type_id
        where p.course_no= #{courseNo}
    </select>
    <!--根据 typeId 查询-->
    <select id="loadCourseByTypeId" parameterType="int" resultType="string">
        select course_no from tbl_course
        where type_id= #{typeId}
    </select>
    <!--带分页查询，注意 MyBatis 中如果填写集合类型，则只填写集合中元素的类型-->
    <select id="loadScopedCourses" parameterType="map" resultType="xiao.ze.
    demo.entity.Course" resultMap="courseRM">
        select * from tbl_course p
        left join tbl_course_type b
        on p.type_id=b.type_id
        <where>
            1=1
            <if test="qryCourseName!=null">and p.course_name like concat(concat
            ('%', #{qryCourseName}), '%')</if>
            <if test="qryStartPoint!=null">and p.course_point >=
            #{qryStartPoint}</if>
            <if test="qryEndPoint!=null">and p.course_point <![CDATA[ <= ]]>
            #{qryEndPoint}</if>
            <if test="typeId!=null">and p.type_id = #{typeId}</if>
        </where>
        order by p.course_no
    </select>
</mapper>
```

文件 UserMapper.xml 的代码如例 10-23 所示。

【例 10-23】 文件 UserMapper.xml 的代码示例。

```
<?xml version="1.0" encoding="UTF-8" ?>
<!DOCTYPE mapper
        PUBLIC "-//mybatis.org//DTD Mapper 3.0//EN"
        "http://mybatis.org/dtd/mybatis-3-mapper.dtd">
<!--配置命名空间，区别名称-->
<mapper namespace="xiao.ze.demo.mapper.UserMapper">
    <!-- SQL 片段-->
    <sql id="cols">user_no,
```

```
        user_name,
        user_pwd
    </sql>
    <resultMap type="xiao.ze.demo.entity.User" id="userRM">
    <!--主键映射-->
    <id property="userNo" column="user_no"/>
    <!--普通字段 property 指实体的属性，column 结果集的字段名称-->
    <result property="userName" column="user_name"/>
    <result property="userPwd" column="user_pwd"/>
    </resultMap>
    <!--新增-->
    <insert id="addUser" parameterType="xiao.ze.demo.entity.User" >
        insert into tbl_users
        (<include refid="cols"/>)
        values
        (#{userNo},#{userPwd},#{userName})
    </insert>
    <!--根据用户名查询-->
    <select id="loadUserByUserName" parameterType="string" resultType=
    "xiao.ze.demo.entity.User" resultMap="userRM">
        select * from tbl_users
        where user_name = #{userName}
    </select>
</mapper>
```

10.2.11 创建 HTML 文件

在目录 resources/templates 下创建文件 footer.html，代码如例 10-24 所示。

【例 10-24】 创建文件 footer.html 的代码示例。

```
<!DOCTYPE html SYSTEM "http://www.thymeleaf.org/dtd/xhtml1-strict-
thymeleaf-4.dtd">
<html xmlns="http://www.w3.org/1999/xhtml" xmlns:th="http://www.thymeleaf.
org">
<body>
<div th:fragment="copy">
    &copy; 2018 参考网上实例
</div>
<div th:fragment="time">
    <a id="time"></a>
    <script th:inline="javascript">
        function setTime(){
            var dt=new Date();
      var arr_week=new Array("星期日","星期一","星期二","星期三","星期四","星期
      五","星期六");
```

```
            var strWeek=arr_week[dt.getDay()];
            var strHour=dt.getHours();
            var strMinutes=dt.getMinutes();
            var strSeconds=dt.getSeconds();
            if (strMinutes<10) strMinutes="0"+strMinutes;
            if (strSeconds<10) strSeconds="0"+strSeconds;
            var strYear=dt.getFullYear()+"年";
            var strMonth=(dt.getMonth()+1)+"月";
            var strDay=dt.getDate()+"日";
            <!--var strTime=strHour+":"+strMinutes+":"+strSeconds;-->
            strTime=strHour+":"+strMinutes;
            time.innerHTML="当前时间为 "+strYear+strMonth+strDay+
            "  "+strTime+"  "+strWeek;
        }
        setInterval("setTime()",1000);
    </script>
</div>
</body>
</html>
```

在目录 resources/templates 下创建文件 index.html，代码如例 10-25 所示。

【例 10-25】 创建文件 index.html 的代码示例。

```
<!DOCTYPE html>
<html lang="en">
<head>
    <meta charset="UTF-8">
    <title>index</title>
    <script type="text/javascript">
        var loginPage = "security/toLogin";
        location.href=loginPage;
    </script>
</head>
<body>
</body>
</html>
```

在目录 resources/templates 下创建文件 login.html，代码如例 10-26 所示。

【例 10-26】 创建文件 login.html 的代码示例。

```
<!DOCTYPE html>
<html lang="en" xmlns:th="http://www.thymeleaf.org">
<head>
    <meta charset="UTF-8">
    <title>课程管理系统登录页面</title>
    <link rel="stylesheet" type="text/css" th:href="@{/css/style.css}" />
    <script type="text/javascript" th:src="@{/js/jquery-3.1.1.min.js}">
```

```
    </script>
</head>
<body>
<div>
    <div id="wrapper" style="text-align:center">
        <img id="login"  width="100" height="20"  th:src="@{/pics/
        logo.png}" /><div id="f_title">课程管理系统用户登录</div><br>
        <form th:action="@{/security/login}" method="post" th:object=
        "${user}">
            <div class="f_row">
                <span>用户姓名:</span>
             <input type="text" class="form-control" name="userName"
             placeholder="请输入姓名">
            </div>
            <div class="f_row">
                <span>登录密码:</span>
              <input type="text" class="form-control" name="userPwd"
              placeholder="请输入密码">
            </div>
            <br>
            <div class="f_row">
                <input type="submit" value=" 登 录 "></input>
                <a th:href="@{/security/register}" class="zcxy" target=
                "_blank">注册</a>
            </div>
        </form>
    </div>
</div>
<div id="footer">
    <div th:include="footer :: copy"></div><div th:include="footer :: time">
    </div>
</div>
</body>
</html>
```

在目录 resources/templates 下创建文件 main.html，代码如例 10-27 所示。

【例 10-27】 创建文件 main.html 的代码示例。

```
<!DOCTYPE html>
<html lang="en" xmlns="http://www.w3.org/1999/xhtml" xmlns:th="http://www.
thymeleaf.org">
<head>
    <meta charset="UTF-8">
    <title>课程管理系统</title>
    <link rel="stylesheet" type="text/css" th:href="@{/css/style.css}" />
</head>
```

```
<body>
<div id="header">
    <div id="productName">
        课程管理系统
     <img id="headpic"  width="30" height="30"  th:src="@{/pics/sbl.
     png}" />
    </div>
    <div style="float:right; margin:30px;">
        操作员:[[${user.userName}]] 
        <a th:href="@{/security/logout}" title="单击离开系统" id="logout">
            <img id="lopics"  width="30" height="30"  th:src="@{/pics/
            logout.jpg}" />
            <span style="cursor: pointer" class="linkspan">退出</span>
        </a>
    </div>
</div>
<div>
    <div id="navigator">
        <div class="menuitem">
            <a th:href="@{/courseType/toInput}" target="contentFrame">新增
            课程类型</a>
        </div>
        <div class="menuitem">
            <a th:href="@{/courseType/list}"  target="contentFrame">课程类型
            管理</a>
        </div>
        <div class="seperator"></div>
        <div class="menuitem">
            <a th:href="@{/course/toInput}"  target="contentFrame">开课登记</a>
        </div>
        <div class="menuitem">
            <a th:href="@{/course/list}" target="contentFrame">课程管理</a>
        </div>
    </div>
    <div id="content">
        <iframe id="contentFrame" width="100%" scrolling="no" height=
        "480px" frameborder="0" name="contentFrame" allowtransparency=
        "true" th:src="@{/welcome.html}">
        </iframe>
    </div>
</div>
<div id="footer">
    <div th:include="footer :: copy"></div><div th:include="footer :: time">
    </div>
</div>
```

```
<script type="text/javascript" th:src="@{/js/jquery-3.1.1.min.js}">
</script>
<script  th:inline="javascript">
    $("#logout").click(function () {
        if(confirm('您真的要离开系统吗？')){
           $(location).attr("href",this.href());
        }
        return false;
    });
</script>
</body>
</html>
```

在目录 resources/templates 下创建文件 register.html，代码如例 10-28 所示。

【例 10-28】 创建文件 register.html 的代码示例。

```
<!DOCTYPE html>
<html lang="en" xmlns:th="http://www.thymeleaf.org">
<!--html lang="en"-->
<head>
    <meta charset="UTF-8" />
    <title>Title</title>
</head>
<body>
<div class="web_login">
    <form name="form2" id="regUser" accept-charset="utf-8"  action=
    "/security/addregister" method="post">
        <ul class="reg_form" id="reg-ul">
            <div id="userCue" class="cue">用户注册，请注意格式</div>
                <label for="username"  class="input-tips2">用户名：</label>
                <div class="inputOuter2">
     <input type="text" id="username" name="username" maxlength="16" class=
     "inputstyle2"/>
                </div>
                <label for="password" class="input-tips2">密码：</label>
                <div class="inputOuter2">
   <input  type="password"  id="password"   name="password"  maxlength="16"
   class="inputstyle2"/>
                </div>
                <label for="password2" class="input-tips2">确认密码：</label>
                <div class="inputOuter2">
   <input type="password" id="password2" name="password2" maxlength="16"
   class="inputstyle2" />
                </div>
                <div class="inputArea">
        <a th:href="@{/security/readdoc}" class="zcxy" target="_blank">
```

```
            阅读注册协议</a>
                    <input type="submit" id="reg"  style="margin-top:10px;
                    margin-left:10px;" class="button_blue" value="同意协议并注册"/>
                </div>
            <div class="cl"></div>
        </ul>
    </form>
</div>
</body>
</html>
```

在目录 resources/templates 下创建文件 readdoc.html，代码如例 10-29 所示。

【例 10-29】 创建文件 readdoc.html 的代码示例。

```
<!DOCTYPE html SYSTEM "http://www.thymeleaf.org/dtd/xhtml1-strict-
thymeleaf-4.dtd">
<html xmlns="http://www.w3.org/1999/xhtml"      xmlns:th="http://www.
thymeleaf.org">
<body>
<div>
    <span>
        注意事项:
        必须注册后才能登录。
    </span><a th:href="@{/security/register}" >返回注册页面</a>
</div>
</body>
</html>
```

在 resources/templates/courceType 下创建文件 input_course_type.html，代码如例 10-30 所示。

【例 10-30】 创建文件 input_course_type.html 的代码示例。

```
<!DOCTYPE html>
<html lang="en" xmlns:th="http://www.thymeleaf.org">
<head>
    <meta charset="UTF-8">
    <title>新增课程类型</title>
    <link rel="stylesheet" type="text/css" th:href="@{/css/style.css}" />
    <script type="text/javascript" th:src="@{/js/jquery-3.1.1.min.js}">
    </script>
</head>
<body style="padding:8px;">
    <h3 class="title">新增课程类型</h3>
    <form th:action="@{/courseType/create}" method="post" th:object=
    "${courseType}">
        <div>
```

```
            <span>课程类型名称:</span>
            <input type="text" name="typeName" placeholder="课程类型名称">
        </div>
        <div>
            <input type="submit" value=" 确定 "/>
        </div>
    </form>
</body>
</html>
```

在目录 resources/templates/courceType 下创建文件 list_course_type.html，代码如例 10-31 所示。

【例 10-31】 创建文件 list_course_type.html 的代码示例。

```
<!DOCTYPE html>
<html lang="en" xmlns:th="http://www.thymeleaf.org">
<head>
    <meta charset="UTF-8">
    <title>课程类型管理</title>
    <link rel="stylesheet" type="text/css" th:href="@{/css/style.css}" />
</head>
<body style="padding:8px;">
<h3 class="title">课程类型管理</h3>
<form action="" method="POST">
    <input type="hidden" name="_method" value="DELETE"/>
</form>
<table border="0" cellspacing="0">
    <tr>
        <th>编号</th>
        <th>名称</th>
        <th>操作</th>
    </tr>
    <tr class="type" th:each="courseType : ${page.list}">
        <td th:text="${courseType.typeId}" nowrap></td>
        <td id="typeName" th:text="${courseType.typeName}" nowrap></td>
        <td nowrap>
<button class="update" th:href="@{/courseType/preUpdate/{typeId}(typeId=
${courseType.typeId})}" >修改</button>
<button class="delete" th:href="@{/courseType/remove/{typeId}(typeId=$
{courseType.typeId})}">删除</button>
        </td>
    </tr>
</table>
<div id="pageInfo">
    共[[${page.total}]]条,
    <span th:if="${ ( page.pageNum - 1 ) * page.pageSize + 1 == page.total }">
```

```
            当前显示第[[${page.total}]]条,
    </span>
    <span th:if="${ ( page.pageNum - 1 ) * page.pageSize + 1 != page.total }">
            当前显示[[${page.startRow}]]-[[${page.endRow}]]条,
    </span>
    第[[${page.pageNum}]]/[[${page.pages}]]页
    |
    <a href="#" th:if="${page.pageNum>1}"><span class="linkspan" id="one">
    首页</span> </a>
    <a href="#" th:if="${page.pageNum>1}"><span class="linkspan" id="two">
    上一页</span> </a>
    <a href="#" th:if="${page.pageNum<page.pages}"><span class="linkspan"
    id="three">下一页</span> </a>
    <a href="#" th:if="${page.pageNum!=page.pages}"><span class="linkspan"
    id="four">末页</span> </a>
    |
到 <input  type="text"  id="pageNo"  size=4  style="text-align:right;"
onkeypress="onlynumber();"/> 页
    <button class="linkspan" id="five" style="color:black;text-
    decoration:none;"> 跳 转 </button>
</div>
<script type="text/javascript" th:src="@{/js/jquery-3.1.1.min.js}">
</script>
<script th:inline="javascript">
    $(function() {
        //删除操作
        $(".delete").click(function () {
            var href = $(this).attr("href");
            if (confirm("确定要删除吗?")) {
                $("form:eq(0)").attr("action", href).submit();
                return false;
            }
        });
        $(".update").click(function () {
            var href = $(this).attr("href");
            $(location).attr("href", href);
        });
        //分页操作
        $(".linkspan").click(function () {
            var pageNo = [[${page.pageNum}]];
            var totalPageNum = [[${page.pages}]];
            var re = /^[0-9]+.?[0-9]*$/;
            if (String($(this).attr("id")) == String("one"))
                pageNo = 1;
            if (String($(this).attr("id")) == String("two"))
```

```
                pageNo = pageNo - 1;
            if (String($(this).attr("id")) == String("three"))
                pageNo = pageNo + 1;
            if (String($(this).attr("id")) == String("four"))
                pageNo = totalPageNum;
            if (String($(this).attr("id")) == String("five")) {
                var num = $.trim($("#pageNo").val());
                if (!re.test(num)) {
                    alert("输入的不是数字!");
                    return;
                }
                pageNo = parseInt(num);
                if (pageNo < 1 || pageNo > totalPageNum) {
                    alert("页号超出范围，有效范围)[1-" + totalPageNum + "]!");
                    return;
                }
            }
        var href = "?pageNo=" + pageNo;
            $(location).attr("href", href);
            return false;
        });
    });
</script>
</body>
</html>
```

在目录 resources/templates/courceType 下创建文件 update_course_type.html，代码如例 10-32 所示。

【例 10-32】 创建文件 update_course_type.html 的代码示例。

```
<!DOCTYPE html>
<html lang="en" xmlns:th="http://www.thymeleaf.org">
<head>
    <meta charset="UTF-8">
    <title>修改课程类型</title>
    <link rel="stylesheet" type="text/css" th:href="@{/css/style.css}" />
    <script type="text/javascript" th:src="@{/js/jquery-3.1.1.min.js}">
    </script>
</head>
<body style="padding:8px;">
    <h3 class="title">修改课程类型</h3>
    <form th:action="@{/courseType/update}" method="post" th:object=
    "${courseType}">
        <input type="hidden" name="_method" value="PUT"/>
        <input type="hidden" name="typeId" th:value="*{typeId}">
        <div>
```

```
            <span>课程类型名称:</span>
 <input type="text" name="typeName" th:value="*{typeName}" placeholder=
 "课程类型名称">
        </div>
        <div>
            <input type="submit" value=" 确定 "/>
        </div>
    </form>
</body>
</html>
```

在目录 resources/templates/cource 下创建文件 input_course.html，代码如例 10-33 所示。

【例 10-33】 创建文件 input_course.html 的代码示例。

```
<!DOCTYPE html>
<html lang="en" xmlns:th="http://www.thymeleaf.org">
<head>
    <meta charset="UTF-8">
    <title>新增课程类型</title>
    <link rel="stylesheet" type="text/css" th:href="@{/css/style.css}" />
</head>
<body style="padding:8px;">
    <h3 class="title">新增课程</h3>
    <img id="textbookPic"
         alt="默认教材封面"
         width="300"
         height="250"
         style="float:right" th:src="@{/pics/SpringBoot.png}" /><br/>
    <form th:action="@{/course/create}" method="post"  enctype="multipart/
    form-data" th:object="${course}">
        <div>
            <span>课程编号:</span>
            <input type="text" name="courseNo" >
        </div>
        <div>
            <span>课程名称:</span>
            <input type="text" name="courseName" >
        </div>
        <div>
            <span>教材封面:</span>
            <input id="coursetextbookpic" type="file" name="coursetextbookpic"
            size="40" />
        </div>
        <div>
            <span>课程课时:</span>
            <input type="text" name="courseHours" >
```

```
        </div>
        <div>
            <span>课程学分:</span>
            <input type="text" name="coursePoint" >
        </div>
        <div>
            <span>课程类型:</span>
            <select name="courseType.typeId">
                <option>=请选择=</option>
                <option th:each="list:${courseTypeList}" th:value=
                "${list.typeId}" th:text="${list.typeName}"></option>
            </select>
        </div>
        <div>
            <span>课程状态:</span>
<input type="radio" name="courseStatus" th:value="O" checked>开放公选
  
            <input type="radio" name="courseStatus" th:value="Z">暂不开放
              
            <input type="radio" name="courseStatus" th:value="C">停止授课
              
        </div>
        <div>
            <span>选课条件:</span>
            <input type="checkbox" name="courseReqs" value="a" />大三以上
            <input type="checkbox" name="courseReqs" value="b" />平均成绩 80 分
            <input type="checkbox" name="courseReqs" value="c" />非本专业学生
            <input type="checkbox" name="courseReqs" value="d" />未拖欠学费
        </div>
        <div>
            <span>备注说明:</span>
            <textarea name="courseMemo" rows="6" cols="60" ></textarea>
        </div>
        <div>
            <input type="submit" value="开设课程"/>
        </div>
    </form>
</body>
<script type="text/javascript" th:src="@{/js/jquery-3.1.1.min.js}">
</script>
<script th:inline="javascript">
    $(function() {
        $("#coursetextbookpic").change(function (e) {
            for (var i = 0; i < e.target.files.length; i++) {
                var file = e.target.files.item(i);
```

```
                var freader = new FileReader();
                freader.readAsDataURL(file);
                freader.onload = function (e) {
                    var src = e.target.result;
                    $("#textbookPic").attr("src", src);
                }
            }
        });
    });
</script>
</html>
```

在目录 resources/templates/cource 下创建文件 list_course.html，代码如例 10-34 所示。

【例 10-34】 创建文件 list_course.html 的代码示例。

```
<!DOCTYPE html>
<html lang="en" xmlns:th="http://www.thymeleaf.org">
<head>
    <meta charset="UTF-8">
    <title>课程类型管理</title>
    <link rel="stylesheet" type="text/css" th:href="@{/css/style.css}" />
</head>
<body style="padding:8px;">
<h3 class="title">课程管理</h3>
<form action="" method="POST">
    <input type="hidden" name="_method" value="DELETE"/>
</form>
<div id="queryArea">
    <form th:action="@{/course/list}" method="post" th:object="${helper}">
        课程名称：<input type="text" name="qryCourseName" />  
        学分范围：<input type="text" name="qryStartPoint" size="6" /> - <input
        type="text" name="qryEndPoint" size="6" />  
        课程类型：
        <select name="qryCourseType">
            <option value="">=请选择=</option>
            <option th:each="list:${courseTypeList}" th:value="${list.
            typeId}" th:text="${list.typeName}"></option>
        </select>
        <input type="submit" value="查询"/>
    </form>
</div>
<table border="0" cellspacing="0">
    <tr>
        <th>序号</th>
        <th>编号</th>
        <th>名称</th>
```

```
            <th>课时</th>
            <th>学分</th>
            <th>类型</th>
            <th>状态</th>
            <th>选课要求</th>
            <th>备注</th>
            <th>操作</th>
        </tr>
        <tr th:each="course,iterStat : ${page.list}" stat>
            <td th:text="${iterStat.index+1}" nowrap></td>
            <td th:text="${course.courseNo}" nowrap></td>
            <td nowrap style="padding-top:10px;">
                [[${course.courseName}]]<br>
                <img width="100" height="50" th:alt="${course.courseName+'的教
                材'}" th:src="@{/course/getPic/{courseNo}(courseNo=
                ${course.courseNo})}"/>
            </td>
            <td th:text="${course.courseHours}" nowrap></td>
            <td th:text="${course.coursePoint}" nowrap></td>
            <td th:text="${course.courseType.typeName}" nowrap></td>
            <td nowrap>
                <span th:if="${course.courseStatus eq 'O'}">开放公选</span>
                <span th:if="${course.courseStatus eq 'Z'}">暂不开放</span>
                <span th:if="${course.courseStatus eq 'C'}">停止授课</span>
            </td>
            <td nowrap>
                <span th:each="req : ${course.courseReqs}">
                    <font th:if="${req eq 'a'}">大三以上</font>
                    <font th:if="${req eq 'b'}">平均成绩 80 分</font>
                    <font th:if="${req eq 'c'}">非本专业学生</font>
                    <font th:if="${req eq 'd'}">未拖欠学费</font>
                </span>
            </td>
            <td th:text="${course.courseMemo}" nowrap></td>
            <td>
                <button class="update" th:href="@{/course/preUpdate/{courseNo}
                (courseNo=${course.courseNo})}">修改</button>
                <button class="delete" th:href="@{/course/remove/{courseNo}
                (courseNo=${course.courseNo})}">删除</button>
            </td>
        </tr>
    </table>
    <div id="pageInfo">
        共[[${page.total}]]条,
        <span th:if="${ ( page.pageNum - 1 ) * page.pageSize + 1 == page.total }">
```

```
            当前显示第[[${page.total}]]条,
    </span>
    <span th:if="${ ( page.pageNum - 1 ) * page.pageSize + 1 != page.total }">
            当前显示[[${page.startRow}]]-[[${page.endRow}]]条,
    </span>
    第[[${page.pageNum}]]/[[${page.pages}]]页
    |
    <a href="#" th:if="${page.pageNum > 1}"><span class="linkspan" id="one">
    首页</span> </a>
    <a href="#" th:if="${page.pageNum > 1}"><span class="linkspan" id="two">
    上一页</span> </a>
    <a href="#" th:if="${page.pageNum < page.pages}"><span class="linkspan"
    id="three">下一页</span> </a>
    <a href="#" th:if="${page.pageNum != page.pages}"><span class=
    "linkspan" id="four">末页</span> </a>
    |
到 <input type="text" id="pageNo" size=4 style="text-align:right;"
onkeypress="onlynumber();"/> 页
    <button class="linkspan" id="five" style="color:black;text-decoration:
    none;"> 跳 转 </button>
</div>
<script type="text/javascript" th:src="@{/js/jquery-3.1.1.min.js}">
</script>
<script th:inline="javascript">
    $(function () {
        //删除操作
        $(".delete").click(function () {
            var href = $(this).attr("href");
            if (confirm("确定要删除吗?")) {
                $("form:eq(0)").attr("action", href).submit();
                return false;
            }
        });
        $(".update").click(function () {
            var href = $(this).attr("href");
            $(location).attr("href", href);
        });
        //分页操作
        $(".linkspan").click(function () {
            var pageNo = [[${page.pageNum}]];
            var totalPageNum = [[${page.pages}]];
            var re = /^[0-9]+.?[0-9]*$/;
            if (String($(this).attr("id")) == String("one"))
                pageNo = 1;
            if (String($(this).attr("id")) == String("two"))
```

```
                pageNo = pageNo - 1;
            if (String($(this).attr("id")) == String("three"))
                pageNo = pageNo + 1;
            if (String($(this).attr("id")) == String("four"))
                pageNo = totalPageNum;
            if (String($(this).attr("id")) == String("five")) {
                var num = $.trim($("#pageNo").val());
                if (!re.test(num)) {
                    alert("输入的不是数字!");
                    return;
                }
                pageNo = parseInt(num);
                if (pageNo < 1 || pageNo > totalPageNum) {
                    alert("页号超出范围，有效范围：[1-" + totalPageNum + "]!");
                    return;
                }
            }
            var act="?pageNo="+pageNo;
            $("form:eq(1)").attr("action",act).submit();
            return false;
        });
    });
</script>
</body>
</html>
```

在目录 resources/templates/cource 下创建文件 update_course.html，代码如例 10-35 所示。

【例 10-35】 创建文件 update_course.html 的代码示例。

```
<!DOCTYPE html>
<html lang="en" xmlns:th="http://www.thymeleaf.org">
<head>
    <meta charset="UTF-8">
    <title>修改课程类型</title>
    <link rel="stylesheet" type="text/css" th:href="@{/css/style.css}" />
    <script type="text/javascript" th:src="@{/js/jquery-3.1.1.min.js}">
    </script>
</head>
<body style="padding:8px;">
<h3 class="title">新增课程</h3>
<img id="textbookPic"
     alt="默认教材封面"
     width="300"
     height="250"
     class="imgShow"
style="float:right" th:src="@{/course/getPic/{courseNo}(courseNo=$
```

```
{course.courseNo})}" />
<br/>
<form th:action="@{/course/update}" method="post"  enctype="multipart/
form-data" th:object="${course}">
    <input type="hidden" name="courseNo" th:value="*{courseNo}" />
    <div>
        <span>课程编号: [[${courseNo}]]</span>
    </div>
    <div>
        <span>课程名称:</span>
        <input type="text" name="courseName" th:value="*{courseName}" >
    </div>
    <div>
        <span>教材封面:</span>
        <input id="coursetextbookpic" type="file" name="coursetextbookpic"
        size="40" />
    </div>
    <div>
        <span>课程课时:</span>
        <input type="text" name="courseHours" th:value="*{courseHours}" >
    </div>
    <div>
        <span>课程学分:</span>
        <input type="text" name="coursePoint" th:value="*{coursePoint}" >
    </div>
    <div>
        <span>课程类型:</span>
        <select name="courseType.typeId" th:value="*{courseType.typeId}">
<option th:each="list:${courseTypeList}" th:value="${list.typeId}"
th:text="${list.typeName}">
</option>
        </select>
    </div>
    <div>
        <span>课程状态:</span>
        <input type="radio" name="courseStatus" th:value="O" th:checked=
        "${course.courseStatus eq 'O'}">开放公选  
        <input type="radio" name="courseStatus" th:value="Z" th:checked=
        "${course.courseStatus eq 'Z'}">暂不开放  
        <input type="radio" name="courseStatus" th:value="C" th:checked=
        "${course.courseStatus eq 'C'}">停止授课  
    </div>
    <div>
        <span>选课条件:</span>
        <input id="a" type="checkbox" name="courseReqs" value="a" />大三以上
```

```
        <input id="b" type="checkbox" name="courseReqs" value="b" />平均成
        绩 80 分
        <input id="c" type="checkbox" name="courseReqs" value="c" />非本专
        业学生
        <input id="d" type="checkbox" name="courseReqs" value="d" />未拖欠
        学费
    </div>
    <div>
        <span>备注说明:</span>
        <textarea name="courseMemo" rows="6" cols="60">[[${course.courseMemo}]]
        </textarea>
    </div>
    <div>
        <input type="submit" value="修改课程"/>
    </div>
</form>
</body>
<script type="text/javascript" th:src="@{/js/jquery-3.1.1.min.js}">
</script>
<script th:inline="javascript">
    $(function() {
        var courseReqs = [[${course.courseReqs}]]
        for (var i = 0; i < courseReqs.length; i++) {
            $("#" + courseReqs[i]).attr("checked", true)
        }
        $('#coursetextbookpic').on('change', function () {
            var file = $(this)
            var fileObj = file[0];  //获取当前元素
            var dataURL;
            var windowURL = window.URL;
            if (fileObj.files[0]) {
                dataURL = windowURL.createObjectURL(fileObj.files[0])
         //创建一个新的对象 URL，该对象 URL 可以代表某一个指定的 File 对象或 Blob 对象
                $('.imgShow').attr('src', dataURL);
            }
            else {
                dataURL = file.val();
                console.log(dataURL)
                $('.imgShow').style.filter = 'progid:DXImageTransform.
                Micsoft.AlphaImageLoader(sizingMethod = scale)'
$('.imgShow').filters.item('DXImageTransform.Microsoft.AlphaImageLoader
').src = dataURL;
            }
            $.ajaxFileUpload({
                url: "",
```

```
                fileElementId: "uploadImg",
                dataType: "string",
                success: function (data) {
                    //图片路径
                    $('.imgShow').attr("src", data);
                }
            });
        });
    });
</script>
</html>
```

10.2.12　修改和创建配置文件

修改目录 resources 下配置文件 application.properties，代码如例 10-36 所示。

【例 10-36】 修改目录 resources 下配置文件 application.properties 的代码示例。

```
#springboot 乱码解决
server.tomcat.uri-encoding=UTF-8
spring.http.encoding.charset=UTF-8
#热部署
spring.devtools.restart.enabled=true
spring.thymeleaf.mode=HTML5
spring.thymeleaf.encoding=UTF-8
spring.thymeleaf.cache=false
spring.thymeleaf.suffix=.html
#加载 mybatis 配置文件
mybatis.mapper-locations=classpath:mybatis/*Mapper.xml
```

在 resources 目录下创建配置文件 application.yml，代码如例 10-37 所示。

【例 10-37】 在 resources 目录下创建配置文件 application.yml 的代码示例。

```
spring:
  datasource:
    url: jdbc:mysql: //localhost:3306/xiaozedb
    driver-class-name: com.mysql.jdbc.Driver
    username: root
    password: sa
druid:
  filters: stat
  maxActive: 20
  initialSize: 1
  maxWait: 30000
  minIdle: 10
  maxIdle: 15
  timeBetweenEvictionRunsMillis: 60000
```

```
  minEvictableIdleTimeMillis: 300000
  validationQuery: SELECT 1
  testWhileIdle: true
  testOnBorrow: false
  testOnReturn: false
  maxOpenPreparedStatements: 20
  removeAbandoned: true
  removeAbandonedTimeout: 1800
  logAbandoned: true
```

10.2.13 创建 CSS 文件

在 resources/static/css 下创建文件 style.css，代码如例 10-38 所示。

【例 10-38】 在 resources/static/css 下创建文件 style.css 的代码示例。

```
body,tr,td,th,div,span,ul,li,input,select,option,textarea{
    font-size:14px;
    font-family:verdana;
    padding:0px;
    margin:0px;
}
#header{
   height:80px;
   border-bottom:2px solid blue;
}
#productName{
    font-size:30px;
    font-weight:bold;
    float:left;
    padding:10px;
    margin-top:20px;
}
#navigator{
    width:100px;
    padding:3px;
    height:500px;
    border-right:2px solid blue;
   float: left;
}
#navigator .menuitem{
   font-size:16px;
   text-align:left;
   margin-top:8px;
}
#navigator .seperator{
```

```
   border-bottom:1px dotted blue;
   margin-top:10px;
   margin-bottom:10px;
}
#content{
   overflow:hidden;
   height:500px;
}
form div {
  margin:5px;
}
div .text{
  border:1px solid blue;
}
table{
     width:100%;
}
table th{
     margin:3px;
     border-bottom:2px solid #cccccc;
     font-weight:bold;
     word-break:keep-all;
}
table tr td{
     text-align:center;
     margin:3px;
     border-bottom:1px dotted #999;
     word-break:keep-all;
}
button{
     margin:3px;
}
.title{
     color:blue;
     font-size:14pt;
    text-decoration:underline;
}
.warnMsg{
     margin:30px;
     height:60px;
     text-align:center;
     border:2px solid yellow;
     color:red;
     font-size:14px;
     padding-top:20px;
```

```
}
#pageInfo{
  text-align:right;
  padding-right:10px;
  font-family:verdana;
  margin-top:3px;
}
.linkspan{
   color:blue;
   text-decoration:underline;
   cursor:hand;
}
#wrapper{
    margin:10px;
}
#f_title{
    font-size:14px;
    font-weight:bold;
    width:100%;
    border-bottom:1px solid gray;
    padding-border:2px;
}
.f_row{
    margin-top:12px;
}
.f_row span,.f_row input{
    font-size:12px;
}
.error{
    border:1px solid yellow;
    color:red;
    width:180px;
    padding:8px;
    margin:5px;
}
#footer{
    font-size:14px;
    text-align:center;
    border-top:2px solid blue;
    margin-top: 1px;
    height: 20px;
    line-height: 20px;
}
```

10.2.14 配置辅助文件并运行程序

下载 jquery-3.1.1.min.js 到 resources/static/js 目录下，在 resources/static/pics 目录下配置 logo.png、sbl.png、logout.png、SpringBoot.png、SpringMVC.png、SSH.png、SSM.png、JavaEE.png 等图片。

在浏览器中输入 localhost:8080，结果如图 10-1 所示。

习题 10

实验题

请独立完成本章的案例。

附录 A

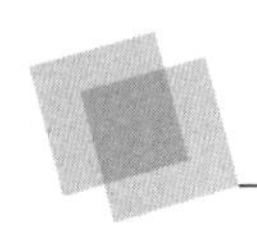

简易天气预报系统的开发

本附录介绍一个简易天气预报系统的开发。

A.1　开发过程

A.1.1　新建项目添加依赖

新建项目 weather，确保在文件 pom.xml 的<dependencies>和</dependencies>之间添加了 Web、Httpcore、Httpclient、Lombok、Quartz、Redis 等依赖。

A.1.2　创建类

依次在包 com.bookcode 下创建 entity、service、controller、config、job、util 等子包，并在包 com.bookcode.entity 中创建类 City、CityList、Forecast、Weather、WeatherResponse、Yesterday，在包 com.bookcode.service 中创建类 CityDataService、WeatherDataService、WeatherReportService，在包 com.bookcode.controller 中创建类 WeatherReportController，在包 com.bookcode.config 中创建类 BlockQuartzConfiguration，在包 com.bookcode.job 中创建类 WeatherDataSyncJob，在包 com.bookcode.util 中创建类 XmlBuilder，并修改它们的代码。

A.1.3　创建文件

在目录 src/main/resources/static 下创建文件 citylist.xml 和 report.js，在目录 src/main/resources/templates 下创建文件 report.html，在目录 src/main/resources 下创建配置文件 application.yml，并修改它们的代码。

上述文件的具体代码请参考本书附带的源代码。完成上述任务后，整个项目的核心目录和部分文件结构如图A-1所示。

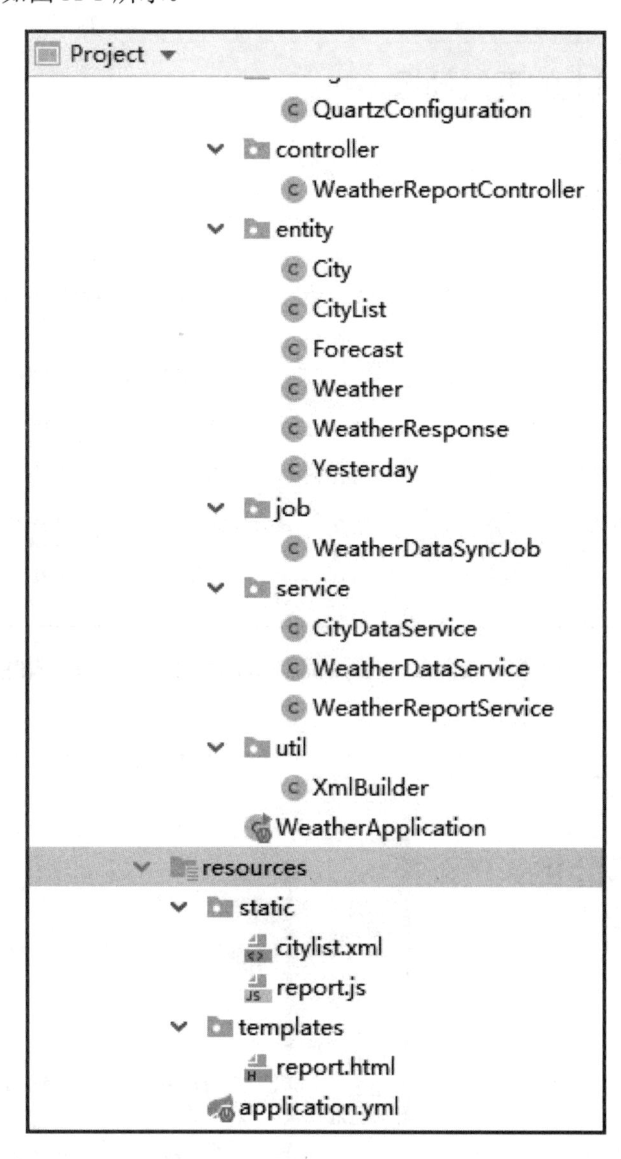

图A-1 项目的核心目录和部分文件结构

A.2 程序结果

运行程序，在浏览器中输入localhost:8080/report/cityId/101190801后结果如图A-2所示。在图A-2的下拉列表框中可以选择其他城市（如南京），如图A-3所示；选择完城市后，自动跳转到所选城市的天气预报页面，如图A-4所示。

与本系统类似的代码可以参考网址https://github.com/oldDong/micro-weather-report，读者可以对比此网址的代码和本书附带的源代码，加深对Spring Boot开发的认识。

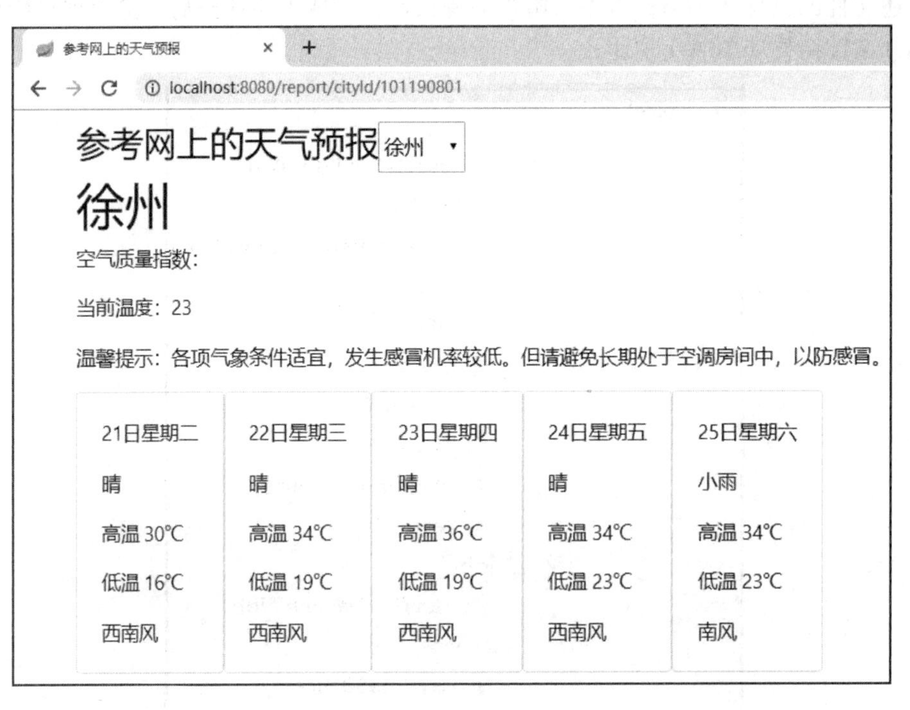

图 A-2 在浏览器中输入 localhost:8080/report/cityId/101190801 后浏览器的输出结果

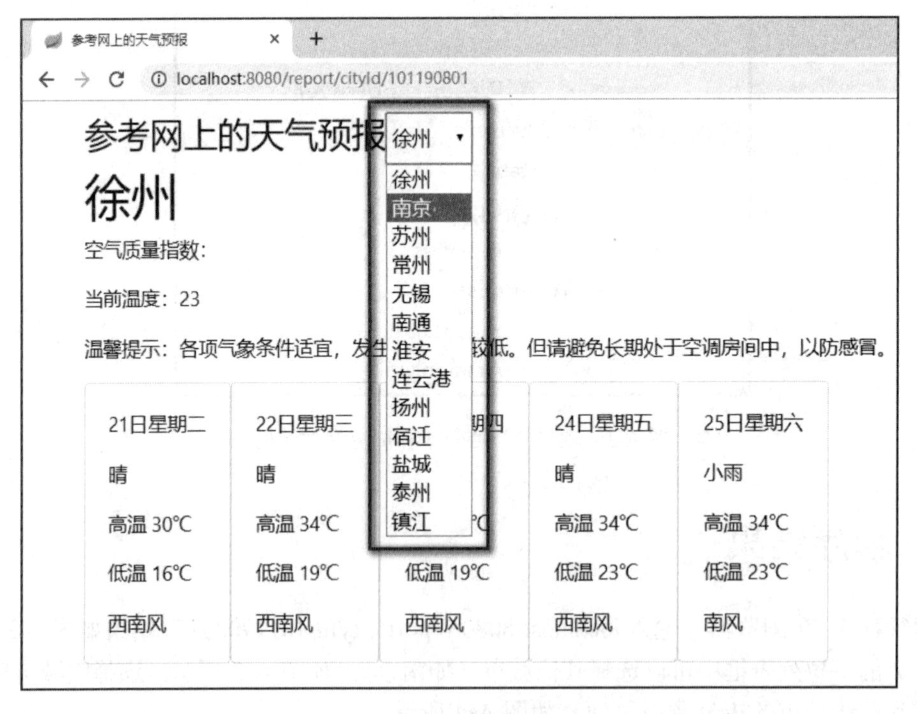

图 A-3 在图 A-2 中下拉列表框中选择要查询天气预报的城市

图 A-4 自动跳转到在图 A-3 中所选城市的天气预报页面

附录 B

简易签到系统的开发

视频讲解

本附录介绍一个简易签到系统的开发。

B.1 开发过程

B.1.1 新建项目添加依赖

新建项目 sign，确保在文件 pom.xml 的<dependencies>和</dependencies>之间添加了 Web、Lombok、Security、Thymeleaf 等依赖。

B.1.2 创建类和接口

依次在包 com.bookcode 下创建 entity、dao、service、controller 等子包，并在包 com.bookcode.entity 中创建类 BaseEntity、Course、CourseStudent、Student、SignRecorder、Sign、Teacher、HttpResult，在包 com.bookcode.dao 中创建接口 CourseRepository、CourseStudentRepository、SignRecorderRepository、SignRepository、StudentRepository、TeacherRepository，在包 com.bookcode.service 中创建类 CourseService、CourseStudentService、SignRecorderService、SignService、StudentService、TeacherService，在包 com.bookcode.controller 中创建类 CourseController、HomeController、StudentController、SignController、TeacherController，并修改它们的代码。

B.1.3 创建与修改文件

在目录 src/main/resources/templates 下创建文件 index.html、add_course.html、add_ student.html、add_course_student.html、add_course_sign.html、add_teacher.html、begin_sign.html 和 end_sign.html，并修改在目录 src/main/resources 下的配置文件 application.properties。

上述文件的具体代码请参考本书附带的源代码。完成上述任务后，整个项目的核心目录和部分文件的结构如图 B-1 所示。

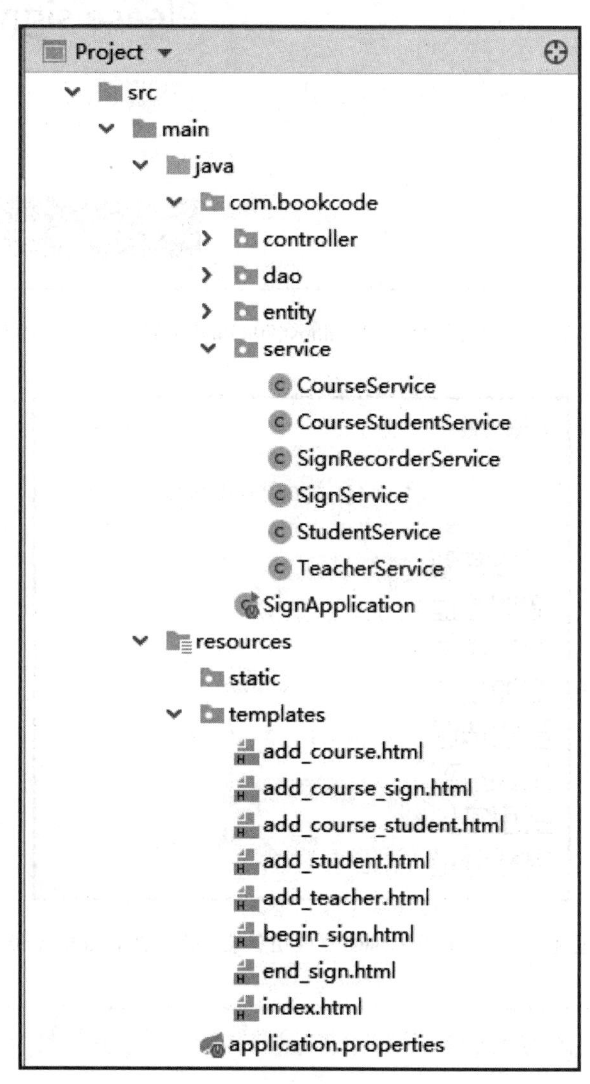

图 B-1　项目的核心目录和部分文件的结构

B.2　程序结果

创建数据库 wechat_sign，运行程序，在浏览器中输入 localhost:8080/index 后跳转到如图 B-2 所示的登录页面。正确输入 Username 和 Password 后单击 Sign in 按钮，登录成功并跳转到 localhost:8080/index 页面，如图 B-3 所示。单击其中的链接，可以重定向到相关的页面，如单击链接“增加课程”后重定向到如图 B-4 所示的界面，在图 B-4 中输入课程相关信息并单击“提交”按钮后，结果如图 B-5 所示。

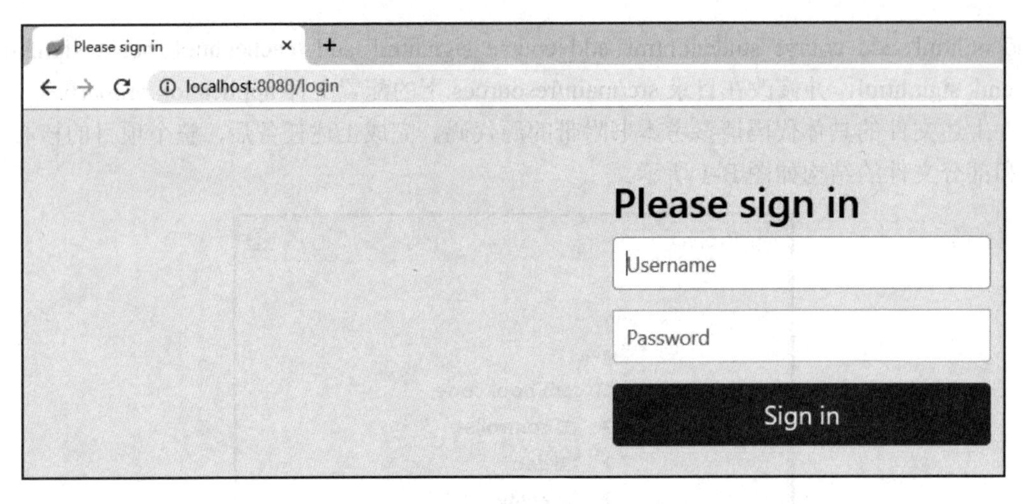

图 B-2　在浏览器中输入 localhost:8080/index 后跳转到登录页面

图 B-3　输入正确 Username 和 Password 成功登录后跳转到 localhost:8080/index 页面

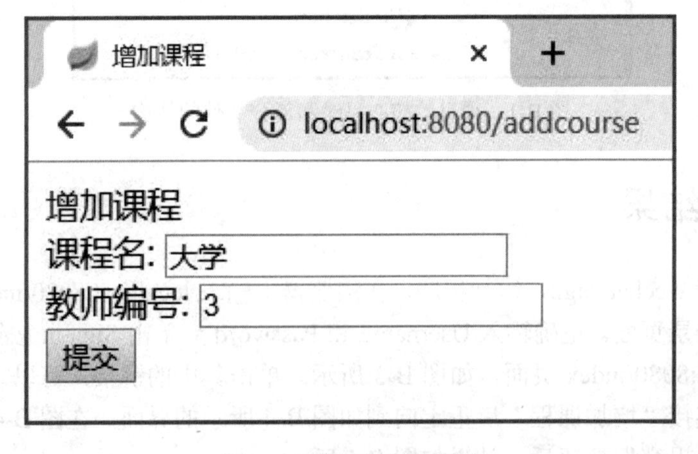

图 B-4　增加课程的界面

图 B-5 增加课程后浏览器的输出结果

与本系统类似的代码可以参考网址 https://github.com/vansl/wechatSign，读者可以对比此网址的代码和本书附带的源代码，加深对 Spring Boot 开发的认识。

附录 C

作为微信小程序后台的简单应用

视频讲解

Spring Boot 因其轻量级的开发方式而受到大家的追捧，这使得它成为微信小程序后台开发比较好的开发工具。本附录介绍 Spring Boot 作为微信小程序后台的一个简单应用，对 Spring Boot 与微信小程序的整合提供一个入门的介绍。若要深入应用还需要对微信小程序和 Spring Boot 开发有较深入的理解。

C.1　作为后台的 Spring Boot 简单应用的开发

C.1.1　新建项目添加依赖

新建项目 h，确保在文件 pom.xml 的<dependencies>和</dependencies>之间添加了 Web、Lombok 等依赖。

C.1.2　创建类和接口

依次在包 com.bookcode 下创建 entity、dao、controller 等子包，并在包 com.bookcode.entity 中创建类 City，在包 com.bookcode.dao 中创建接口 CityRepository，在包 com.bookcode.controller 中创建类 CityController、HelloController，并修改它们的代码。

C.1.3　修改文件

修改在目录 src/main/resources 下的配置文件 application.properties。

上述文件的具体代码请参考本书附带的源代码。

完成上述任务后，整个项目的核心目录和部分文件结构如图 C-1 所示。

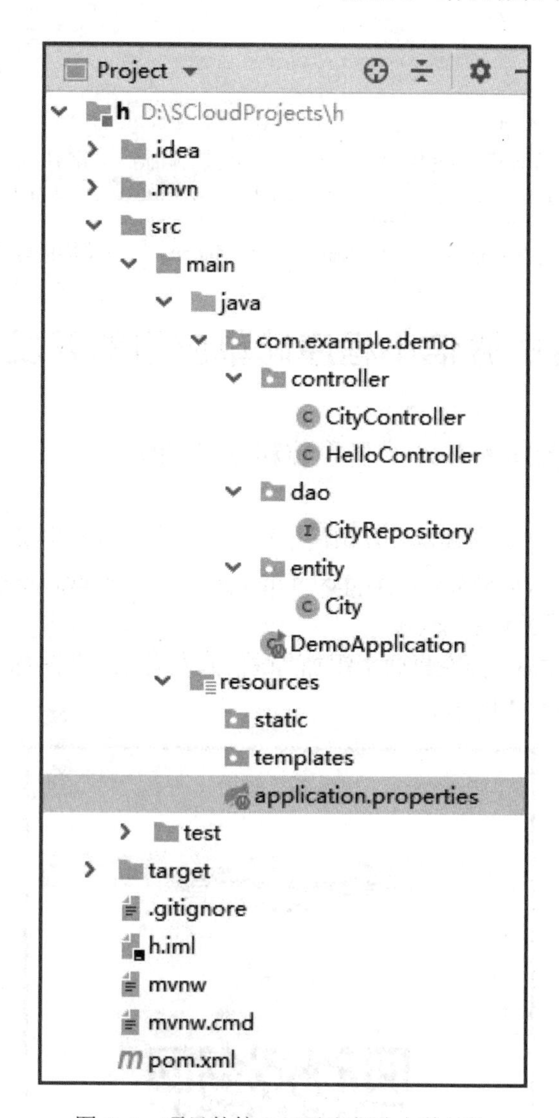

图 C-1　项目的核心目录和部分文件结构

C.1.4　在浏览器中直接访问的结果

运行程序，在浏览器中输入 localhost:8080 后结果如图 C-2 所示。在浏览器中输入 localhost:8080/listcity 后结果如图 C-3 所示。

图 C-2　在浏览器中输入 localhost:8080 后浏览器的输出结果

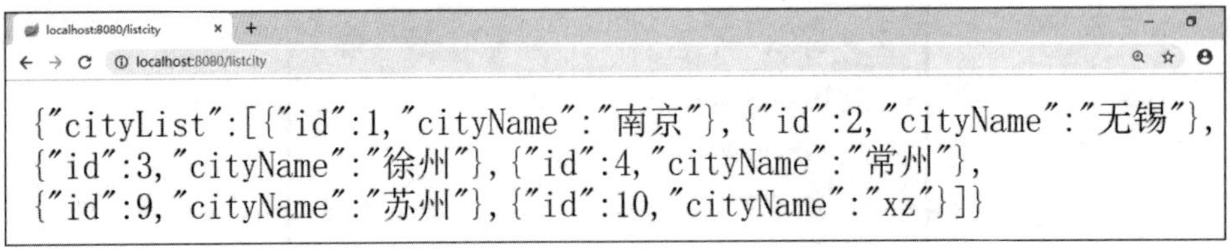

localhost:8080/listcity

```
{"cityList":[{"id":1,"cityName":"南京"},{"id":2,"cityName":"无锡"},
{"id":3,"cityName":"徐州"},{"id":4,"cityName":"常州"},
{"id":9,"cityName":"苏州"},{"id":10,"cityName":"xz"}]}
```

图 C-3　在浏览器中输入 localhost:8080/listcity 后浏览器的输出结果

C.2　作为前台的微信小程序简单应用的开发

C.2.1　微信小程序的开发工具安装和项目创建

为了帮助开发者简单、高效地开发微信小程序，微信官方推出了小程序开发工具。可以从官方网站（https://developers.weixin.qq.com/miniprogram/dev/devtools/download.html）下载该开发工具，并按照导航逐步安装。

开发工具安装完成后，会在桌面上添加“微信开发者工具”图标。双击该图标打开“微信开发者工具”，如图 C-4 所示。

图 C-4　打开“微信开发者工具”

用手机微信扫描二维码，通过验证后显示“扫描成功”，如图 C-5 所示。显示该工具可以用来开发“小程序项目”“公众号网页项目”，如图 C-6 所示。选择“小程序项目”后可以创建一个微信小程序项目。微信小程序开发工具的安装和项目创建的细节本书不做介绍，读者可以参考编者编写的《微信小程序开发基础》一书或其他资料。

图 C-5　通过验证后的显示“扫描成功”

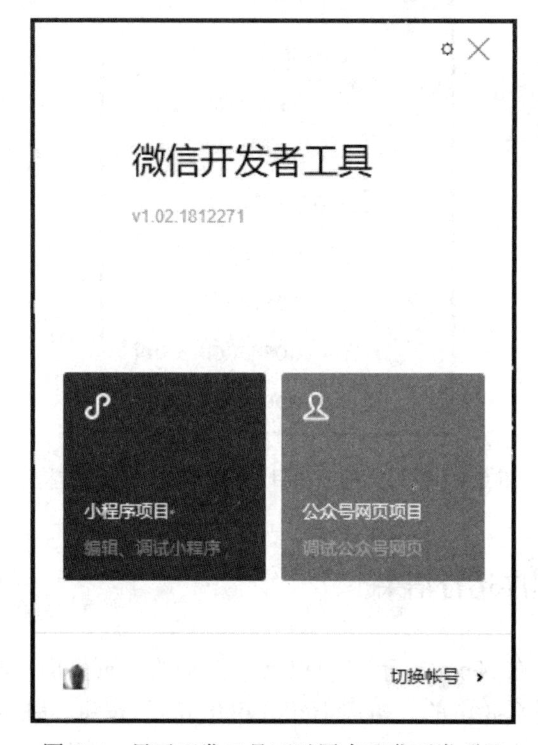

图 C-6　显示开发工具可以用来开发两类项目

C.2.2 新建文件

依次在目录 pages 文件夹中添加 hi、hello、list、operation 4 组文件，每组包括 4 个文件，如 hello.js、hello.json、hello.wxml、hello.wxss 为一组文件。项目增加的目录、文件结构如图 C-7 所示。这些文件的具体代码请参考本书附带的源代码。

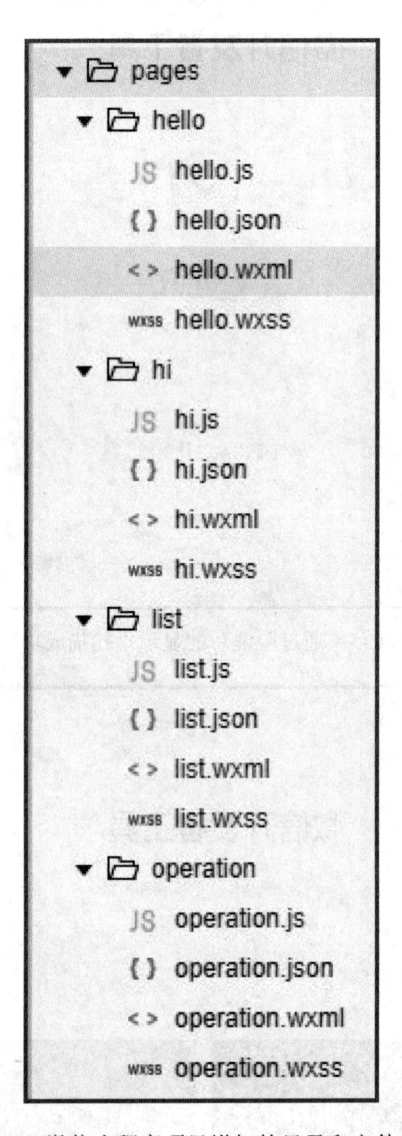

图 C-7　微信小程序项目增加的目录和文件结构

C.2.3 微信小程序运行结果

编译微信小程序，在 Nexus 6 手机模拟器中显示的首页界面（与 hi.wxml 文件对应）如图 C-8 所示。单击图 C-8 中的“访问 HelloController”按钮，跳转到如图 C-9 所示的界面（与 hello.wxml 文件对应）。单击图 C-9 中“访问 http://localhost:8080 的结果”按钮，在开发工具的控制台中输出访问后台 Spring Boot 得到的内容，如图 C-10 所示。对比图 C-10 和图 C-2，可以发现两者的内容一致。单击图 C-8 中的“调用 CityController”按钮，跳转到如图 C-11 所示的界面（与 list.wxml 文件对应）。对比图 C-11 和图 C-3，可以发现两者的内容一致。单击图 C-11 中“添加城市”按钮，跳转到如图 C-12 所示的界面（与 operation.wxml 文件对应）。在图 C-12 的文本框中输入要添加的城市名称（如“北京”）后单击“提交”按钮，结果如图 C-13 所示。再次在浏览器中输入 localhost:8080/listcity 后，结果如图 C-14 所示。对比图 C-13 和图 C-14，可以发现两者的内容一致，这说明前台微信小程序的操作和后台 Spring Boot 进行了关联，并将操作结果存储到了 MySQL 数据库中。

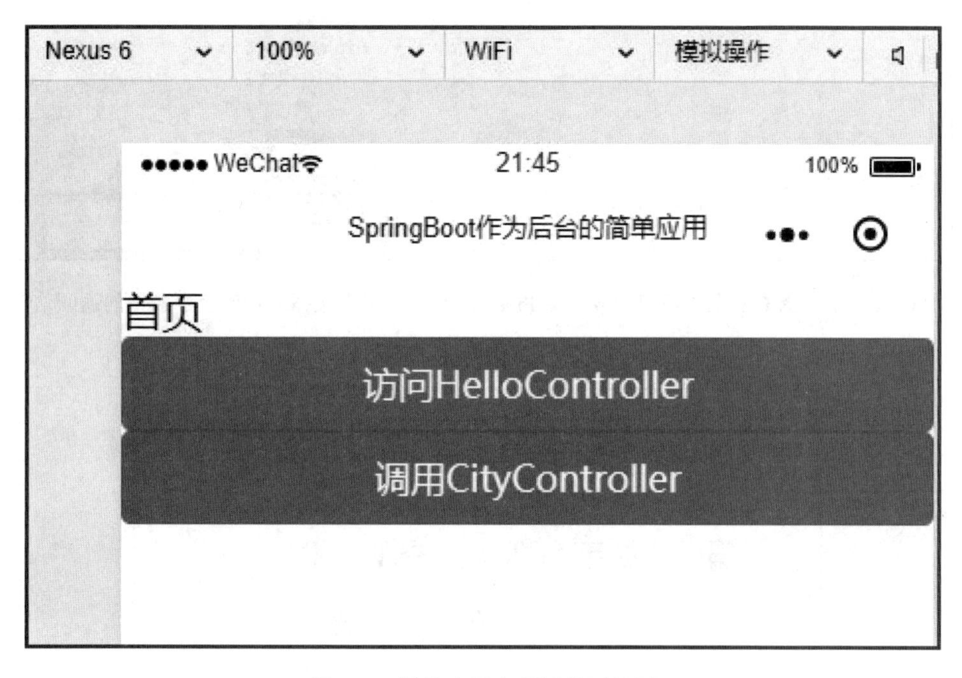

图 C-8　微信小程序项目首页界面

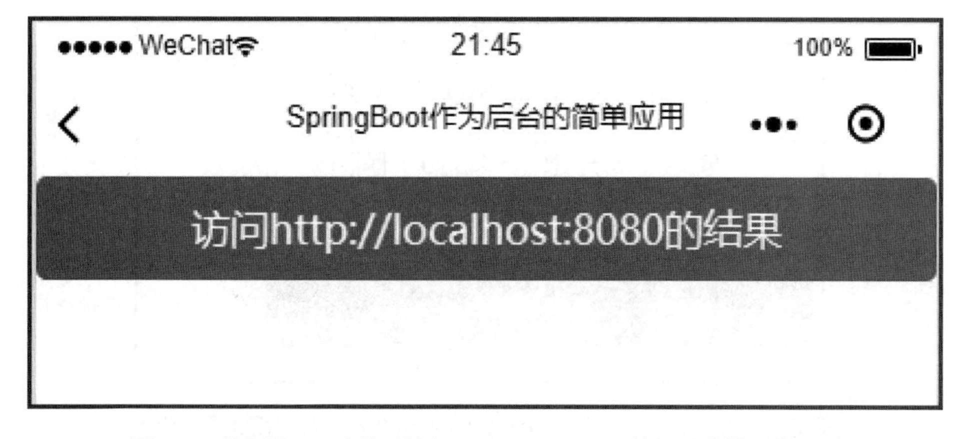

图 C-9　单击图 C-8 中的“访问 HelloController”按钮后跳转到的界面

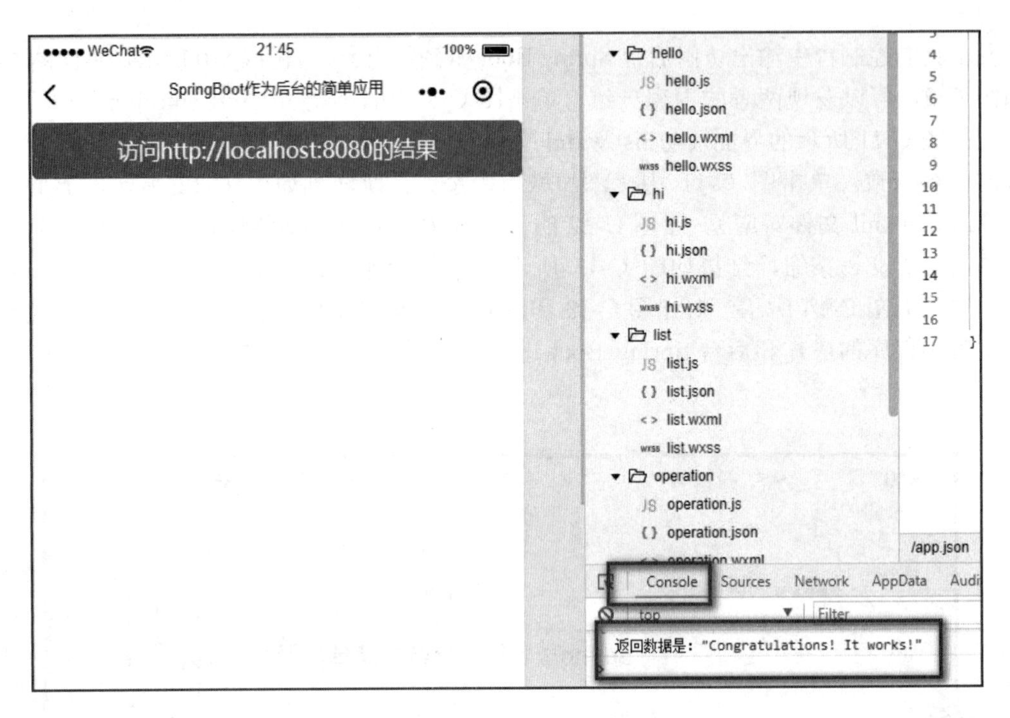

图 C-10 单击图 C-9 中“访问 http://localhost:8080 的结果”按钮后控制台中输出的返回信息

图 C-11 单击图 C-8 中的“调用 CityController”按钮后跳转到的界面

图 C-12　单击图 C-11 中“添加城市”按钮后跳转到的界面

图 C-13　在图 C-12 文本框中输入要添加的城市名称后单击“提交”按钮的结果

```
{"cityList":[{"id":1,"cityName":"南京"},{"id":2,"cityName":"无锡"},
{"id":3,"cityName":"徐州"},{"id":4,"cityName":"常州"},
{"id":9,"cityName":"苏州"},{"id":10,"cityName":"xz"},
{"id":11,"cityName":"北京"}]}
```

图 C-14　再次在浏览器中输入 localhost:8080/listcity 后的结果

C.3　Spring Boot 和微信小程序整合的关键点

C.3.1　两者关联的关键代码

Spring Boot 和微信小程序整合的关键是在微信小程序中访问 Spring Boot 的服务，两者关联的关键代码如例 C-1 所示。

【例 C-1】 两者关联的核心代码示例。

```
//pages/hello/hello.js
...
sayhello: function (e) {
    wx.request({
      url: 'http://localhost:8080/',
      method: 'GET',
      data: {},
      success: function (res) {
        console.log("返回数据是: "+JSON.stringify(res.data));
      }
    })
  },
...
```

C.3.2　注意事项

微信小程序和服务器进行网络通信的方式包括 HTTPS 协议（请注意不是 HTTP 协议）和 WebSocket 等。默认情况下，微信小程序后台只接受 HTTPS 域名，开发时可以申请此类域名；或者在开发工具中进行设置后使用 HTTP 域名。设置方法是：单击工具中“详情”按钮后勾选“不校验合法域名、web-view（业务域名）、TLS 版本以及 HTTPS 证书”前的复选框；勾选后微信小程序就可以使用访问 HTTP 域名（如 localhost:8080）。

与本示例类似的代码可以参考网址 https://blog.csdn.net/sinat_25295611/article/details/79611316，读者可以对比此网址的代码和本书附带的源代码，加深对 Spring Boot 和微信小程序整合开发的认识。

附录 D

Spring Boot 和 Vue.js 的整合开发

视频讲解

Vue.js 是构建用户界面的渐进式框架。与其他重量级框架不同的是，Vue.js 采用自底向上增量开发的设计。Vue.js 的核心库只关注视图层，并且非常容易学习，非常容易与其他库或已有项目整合。本附录介绍 Spring Boot 整合 Vue.js 开发一个简单应用，对 Spring Boot 与 Vue.js 的整合开发提供入门介绍。若要开发复杂应用还需要对 Vue.js 开发和 Spring Boot 开发有较深入的理解。

D.1 Spring Boot 和 Vue.js 的整合

D.1.1 Vue.js 的安装

安装 Vue.js 之前先要安装 Node.js，到 Node.js 官方网站（https://nodejs.org/en/）下载 Node.js 后安装 Node.js。安装 Node.js 时就已经自带了包管理器 npm。安装完 Node.js 之后，打开 Windows 命令处理程序 CMD，依次执行如例 D-1 所示的一系列命令。

【例 D-1】 一系列命令的示例。

```
node  -v     #1 检验 Node.js 是否安装成功
npm  -v      #2 得到 npm 的版本
npm  install  -g  cnpm  --registry=https://registry.npm.taobao.org  #3 安装 cnpm
cnpm  install  -g  vue-cli    #4 安装脚手架 2.X 版，安装 3.X 版用@vue/cli
cd  d:      #5
d:          #6
md  workspace       #7 在 d 盘新建目录 workspace，您也可以改成其他名字
cd  d:\workspace  #8
vue  init  webpack  sb-vue   #9 创建一个项目 sb-vue，您也可以改为其他名称
```

```
cd  d:\workspace\sb-vue   #10
npm  run  dev    #11 运行项目
```

例 D-1 命令后 # 号表示注释，在输入命令时不用输入#及其后面的内容。#后面的数字表示命令的序号，例如#1 表示第 1 条命令；数字后面的文字说明命令的作用。在两条命令之间，前一条命令执行完成之后，才执行后一条命令。如第 1 条命令执行完成之后才能执行第 2 条命令，如图 D-1 所示。

图 D-1　在 Windows 命令处理程序 CMD 中执行完第 1 条命令后再执行第 2 条命令

在执行第 9 条命令时，除了要给出作者的姓名（可以根据需要命名）之外，其余每项选择 Y 后按 Enter 键即可。

正确执行完例 D-1 中命令之后，在浏览器中输入 localhost:8080，结果如图 D-2 所示。安装过程还可以参考网址 https://segmentfault.com/a/1190000013950461。

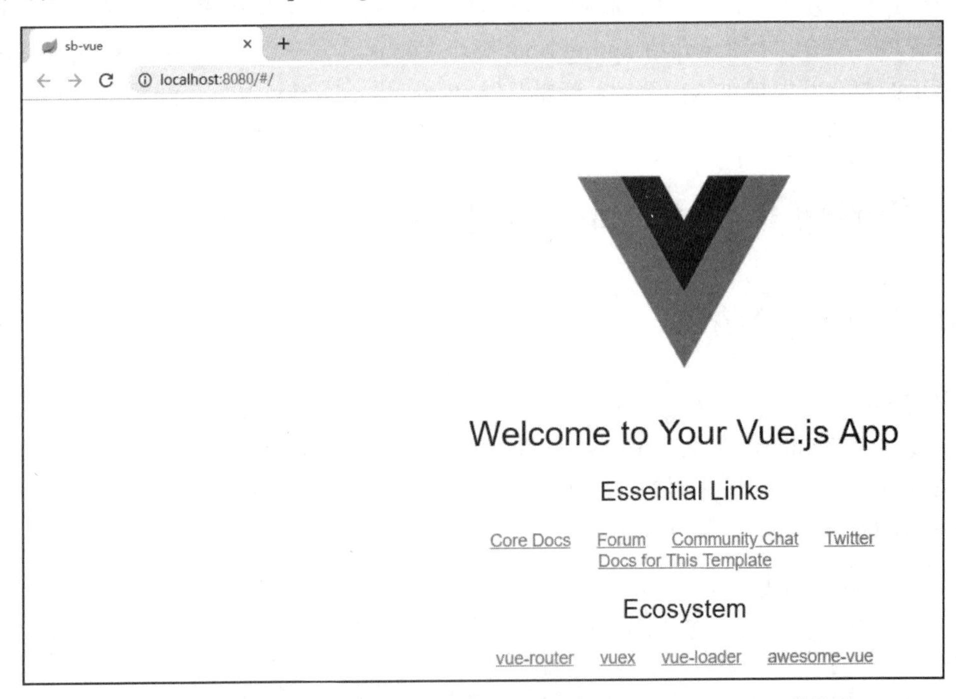

图 D-2　正确执行完例 D-1 中命令后在浏览器中输入 localhost:8080 的结果

D.1.2　在 IDEA 中集成 Vue.js

本书用的集成开发环境是 IDEA，安装完 Vue.js 之后可以将其集成到 IDEA 中。打开

IDEA 的 Plugins 窗口，在查询框中输入 Vue.js，查询到 Vue.js 插件。单击 Vue.js 插件的图标和介绍文字下方的 Install 按钮，IDEA 自动安装 Vue.js 插件，安装结束后 Install 按钮变成 Restart IDE 按钮，如图 D-3 所示。单击 Restart IDE 按钮，重新启动 IDEA；再次打开 IDEA 的 Plugins 窗口并在查询框中输入 Vue.js，Install 按钮变成 Installed 按钮，且按钮颜色变成灰色，显示已经成功安装 Vue.js 插件，如图 D-4 所示。打开 IDEA 后单击 Create New Project 并选择 Static Web 项目，可以看到已有 Vue.js 类型项目可供选择，如图 D-5 所示，说明成功地在 IDEA 中集成了 Vue.js。为了有效使用 Vue.js 还要将 JavaScipt 的版本设置成 ECMAScript6，如图 D-6 所示。

在 IDEA 中集成 Vue.js 的过程还可以参考网址 https://www.jianshu.com/p/d08043088340 和 https://blog.csdn.net/Neuf_Soleil/article/details/88926242。

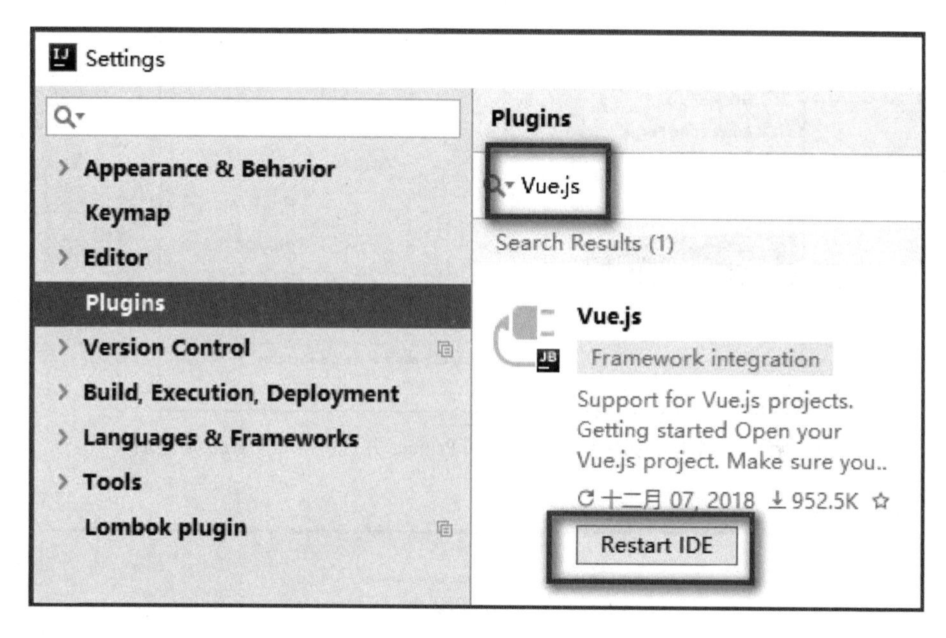

图 D-3　在 IDEA 中安装完 Vue.js 插件

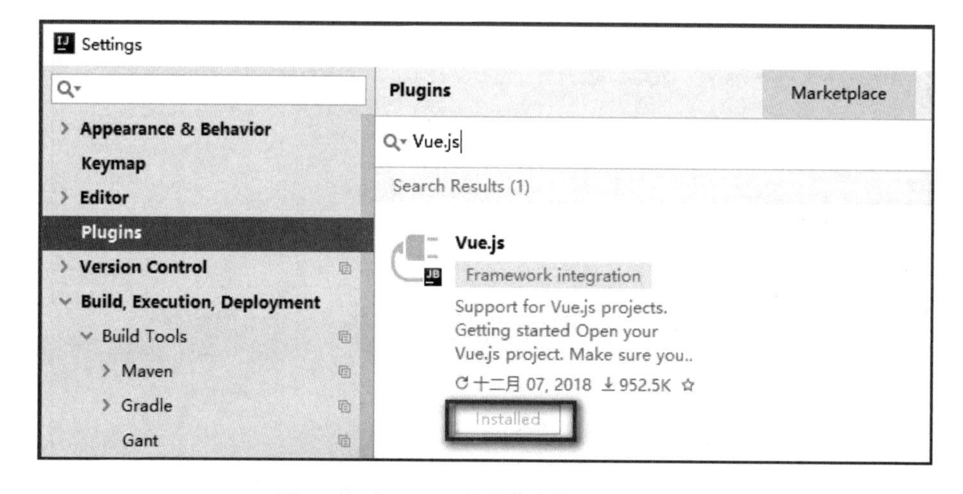

图 D-4　在 IDEA 中成功安装 Vue.js 插件

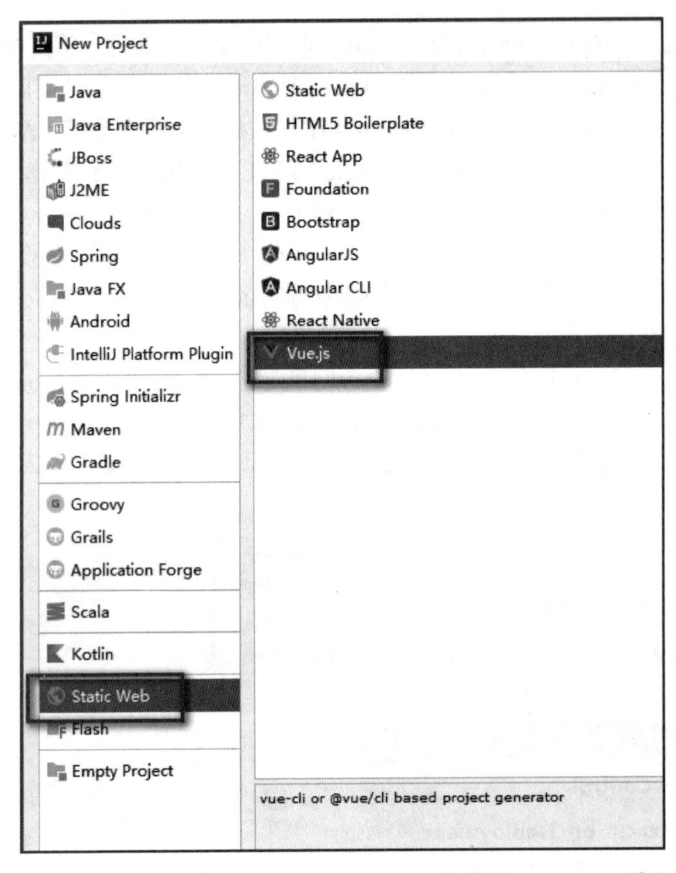

图 D-5　在 IDEA 中单击 Create New Project 并选择 Static Web 项目后可见到 Vue.js 项目的界面

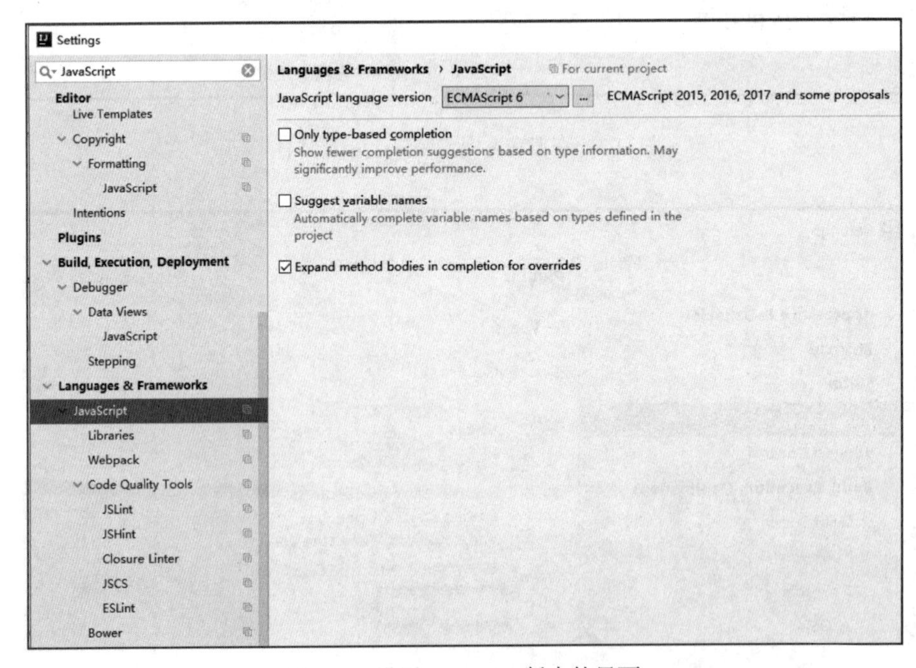

图 D-6　设置 JavaScipt 版本的界面

D.2　Spring Boot 和 Vue.js 整合开发的简单示例

D.2.1　创建 Vue.js 项目 helloworld

打开 IDEA 后单击 Create New Project 并选择 Static Web 项目中 Vue.js 类型项目，如图 D-5 所示。单击 Next 按钮后，将项目名称（Project name）设为 helloworld，将项目位置（Project location）设为 D:\vue-workspace\helloworld，如图 D-7 所示。之后，一直单击 Next 按钮（其中需要设置作者的姓名）即可完成项目的创建，创建完的项目 helloworld 的主要目录和部分文件构成如图 D-8 所示。

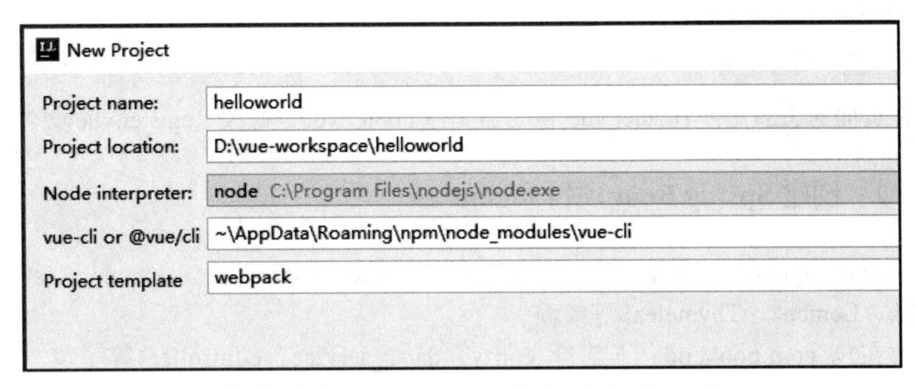

图 D-7　设置项目名称（Project name）和项目位置（Project location）

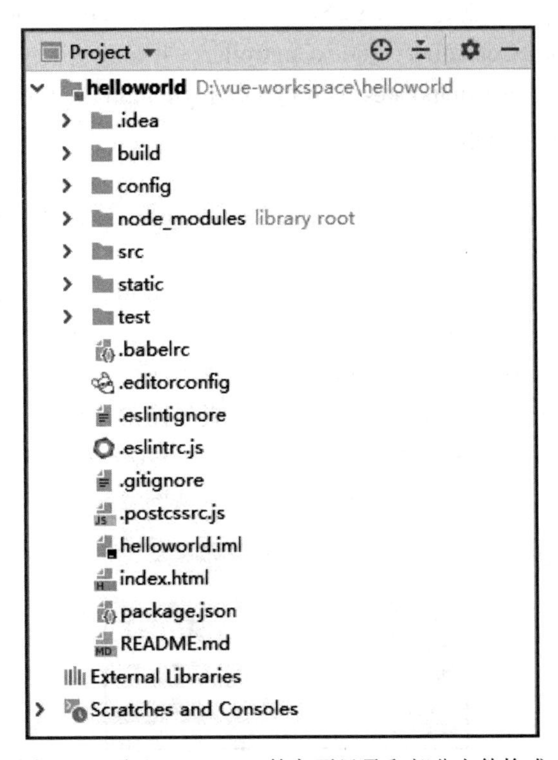

图 D-8　项目 helloworld 的主要目录和部分文件构成

iView 是一套基于 Vue.js 的高质量 UI 组件库，利用它可以更高效地开发界面。打开 Windows 命令处理程序 CMD，执行如例 D-2 所示的命令来安装 iView。

【例 D-2】 安装 iView 的命令示例。

```
npm install --save iview
```

在 main.js 中加入 iView 相关的代码。在项目目录 components 下新建一个 Register.vue 文件，到 iView 的官方网站（https://www.iviewui.com/docs/guide/install）找到 Form 表单组件的代码并复制到 Register.vue 中。编辑完 Register.vue 后修改文件 App.vue。

Vue.js 官方推荐使用 axios 来发送异步请求。打开 Windows 命令处理程序 CMD，执行如例 D-3 所示的命令安装 axios。安装完 axios 后在 main.js 中加入相关代码使用 axios。

【例 D-3】 安装 axios 的命令示例。

```
npm install axios -S
```

设计页面的头部文件 Header.vue 和底部文件 Footer.vue，修改 Register.vue 文件。

D.2.2 创建 Spring Boot 项目 loginuser

新建项目 loginuser，确保在文件 pom.xml 的<dependencies>和</dependencies>之间添加了 Web、Lombok、Thymeleaf 等依赖。

依次在包 com.bookcode 下创建 entity、dao、service、controller 等子包，并在包 com.bookcode.entity 中创建类 User，在包 com.bookcode.dao 中创建接口 UserRepository，在包 com.bookcode.service 中创建类 UserService，在包 com.bookcode.controller 中创建类 RegisterController、HomeController、IndexController，并修改它们的代码。

在目录 src/main/resources/templates 下创建文件 login.html，修改在目录 src/main/resources 下的配置文件 application.properties。

D.2.3 运行结果

为了把两者整合到一起，修改文件 Register.vue 和 config 下的 index.js 等文件。上述文件的具体代码请参考本书附带的源代码。

依次运行项目 loginuser 和 helloworld，在浏览器中输入 localhost:8081/index 的结果如图 D-9 所示，在浏览器中输入 localhost:8080 的结果如图 D-10 所示。单击图 D-10 中链接“跳转到服务器”后，结果如图 D-9 所示。

图 D-9 在浏览器中输入 localhost:8081/index 的结果

图 D-10　在浏览器中输入 localhost:8080 的结果

与本系统类似的代码可以参考网址 https://www.jianshu.com/p/bbc455d86a22 和 https://segmentfault.com/a/1190000014211773，读者可以对比这两个网址的代码和本书附带的源代码，加深对 Spring Boot 和 Vue.js 整合开发的认识。

参 考 文 献

[1] 杨恩雄. Spring Boot 2+Thymeleaf 企业应用实战[M]. 北京：电子工业出版社，2018.

[2] 疯狂软件. Spring Boot 2 企业应用实战[M]. 北京：电子工业出版社，2018.

[3] 陈光剑. Spring Boot 开发实战[M]. 北京：机械工业出版社，2018.

[4] 陈韶健. 深入实践 Spring Boot 开发实战[M]. 北京：机械工业出版社，2016.

[5] 王福强. SpringBoot 揭秘：快速构建微服务体系[M]. 北京：机械工业出版社，2016.

[6] 柳伟卫. Spring Boot 企业级应用开发实战[M]. 北京：北京大学出版社，2018.

[7] 汪云飞. JavaEE 开发的颠覆者：Spring Boot 实战[M]. 北京：电子工业出版社，2016.

[8] Craig Walls. Spring Boot 实战[M]. 丁雪丰，译. 北京：人民邮电出版社，2016.

[9] 李家智. Spring Boot 2 精髓：从构建小系统到架构分布式大系统[M]. 北京：电子工业出版社，2017.

[10] 杨开振. 深入浅出 Spring Boot 2.x[M]. 北京：人民邮电出版社，2018.

图书资源支持

感谢您一直以来对清华版图书的支持和爱护。为了配合本书的使用，本书提供配套的资源，有需求的读者请扫描下方的“书圈”微信公众号二维码，在图书专区下载，也可以拨打电话或发送电子邮件咨询。

如果您在使用本书的过程中遇到了什么问题，或者有相关图书出版计划，也请您发邮件告诉我们，以便我们更好地为您服务。

我们的联系方式：

地　　址：北京市海淀区双清路学研大厦 A 座 701

邮　　编：100084

电　　话：010－62770175－4608

资源下载：http://www.tup.com.cn

客服邮箱：tupjsj@vip.163.com

QQ：2301891038（请写明您的单位和姓名）

资源下载、样书申请

书圈

扫一扫，获取最新目录

用微信扫一扫右边的二维码，即可关注清华大学出版社公众号“书圈”。